Geometry Made Easy Handbook

By:
Mary Ann Casey
B. S. Mathematics, M. S. Education

Acknowledgements

Thank you to my colleague and friend, Kimberly Knisell, for proofreading and making suggestions. Also thank you to my daughter, Debra Casey, who also teaches high school mathematics, for straightening up my technical details. In reference to the line segment symbols, often called hats, my daughter said, "Mom, you are overdoing the hats!" I was overdoing the degree symbols also, but we fixed it all up. Certainly, special thanks to Keith Williams, and his assistant Julieen Kane. Working with math symbols and diagrams is difficult even for a math teacher and these two people put it all together in an attractive and legible format for publication.

Dedication

To my former students who have become math teachers themselves!

About The Author

The author has a B.S. in Mathematics and a M.S. in Education from the state University of New York at New Paltz. She has been teaching high school math for 20 years and has taught several courses at Ulster County Community College. Mrs. Casey is currently the lead math teacher in her high school, has served on committees at the NY State Education Dept., has been named in Who's Who Among American Teachers many times in her career, and maintains professional memberships in the National Council of Teachers of Mathematics and in the Association of Mathematics Teachers in New York State.

Introduction

<u>Geometry Made Easy</u> is an informal quick reference guide that uses student friendly language and techniques. It is coordinated with the New York State Geometry curriculum. In the interest of good teaching of mathematics, some additional information is included to enhance the understanding of the material. This book does not contain practice problems and it is suggested that it can be used in conjunction with Topical's Geometry Regents Practice Test Book (ISBN 978-1-929099-38-2).

Geometry students frequently complain that they don't understand why they have to *write* out things in math. They are used to dealing with numbers and with variables in math, but now suddenly they have to find and *communicate* a reason for each step they take in reaching a mathematical conclusion. The step by step analysis of problem solving is evident through-out the study of geometry and carries on into future mathematical studies. The analytical thinking process and the skill needed to clearly express the reasoning steps used in geometry are invaluable life skills as well, but it is hard to convince students of that!

Each teacher works from his/her own knowledge and teaches specific methods that the students are expected to use. Before writing this handbook I supplemented my own knowledge by reading many resource materials. It was quickly evident that the variety of interpretations used in the study of geometry was very wide -- especially in the classification of definitions, theorems, axioms, etc. As a result, in putting together the *Handbook*, I have used a very informal classification of theorems, etc. as "reasons." It should be clearly understood that each student should use this booklet in close association with the directives of his/her own teacher and textbook.

Sincerely,

MaryAnn Casey
High School Math Teacher

GEOMETRY MADE EASY

Table of Contents

1 — 3 Dimensional Geometric Relationships

The geometric relationships studied in this course include those involving more than one plane, lines not in a plane, and figures like prisms, cylinders, pyramids, and spheres.

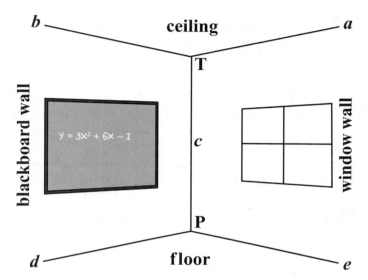

Note: The walls are perpendicular to the floor, the ceiling, and each other.

Relationships between lines and planes:

* If two non-coplanar lines do not intersect, they are skew lines.

  Example Line b and line e are skew lines.

* If a line is perpendicular to each of two intersecting lines at their point of intersection, then the line is perpendicular to the plane containing the intersecting lines.

  Example Line c is perpendicular to lines d and e at P. Therefore line c is perpendicular to the floor.

* Through a given point there is one and only one plane passing through the point that is perpendicular to a given line.

  Example Line c is perpendicular to the floor at P. The floor is the only plane that can go through P that is to perpendicular c.

- Through a given point there is one and only one line perpendicular to a given plane.

  Example Through point P there is only one line, c, that is perpendicular to the floor.

- If two lines are each perpendicular to the same plane, they are coplanar.

  Example Line a and line e are both perpendicular to the blackboard wall. a and e are coplaner.

- Two planes are perpendicular to each other if and only if one plane contains a line perpendicular to the second plane.

  Example Line b is perpendicular to the window wall. Therefore the blackboard wall is perpendicular to the window wall.

- If a line is perpendicular to a plane, then any line perpendicular to the given line at its point of intersection with the given plane is in the given plane.

  Example Line d is perpendicular to the window wall. Therefore line C is in the window wall.

- If a line is perpendicular to a plane, they every plane containing the line is perpendicular to the given plane.

  Example The c is in both walls and is perpendicular to the floor. Therefore the window wall and blackboard wall are both perpendicular to the floor.

- If a plane intersects two parallel planes, then the intersection is two parallel lines.

  Example Since the ceiling and the floor are parallel, the lines, a and e, formed by the intersection of the window wall are parallel to each other.

- If two planes are perpendicular to the same line, the planes are parallel.

  Example The ceiling and the floor are each perpendicular to line c. Therefore the the ceiling and the floor are parallel.

Relationship between spheres and planes:

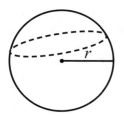

- When a plane intersects a
  sphere, a circle is formed.

- **Great Circle:** If a plane intersects a
  sphere and passes through its center,
  the intersection of the sphere and the
  plane is a great circle. A great circle
  is the largest circle that can be drawn
  on a sphere. It has the same radius as
  the sphere.

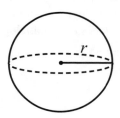

- If two planes intersect a circle at
  equal distances from the center,
  two congruent circles are formed
  by the intersection. They have equal radii.

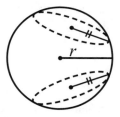

3 dimensional figures and their properties:

Solid: a closed surface, usually including its interior space.

Polyhedron: A closed figure with faces that are planar (flat). The faces meet on line segments called edges and the edges meet at vertices.

- **Prism:** A polyhedron with parallel congruent bases which are both
  polygons. The lateral faces of a prism are all parallelograms or
  rectangles. The lateral edges of a prism are congruent and parallel. The
  height of a prism is the perpendicular distance between the parallel bases.

    Cube: All its faces are congruent
    squares. (Ex: a die)

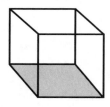

Rectangular Solid: All its faces are rectangles. (Ex: A shoebox)

Parallelepiped: The parallel bases are parallelograms. In this prism, all the faces are parallelograms.

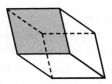

Triangular Prism: The 2 parallel bases are triangles, the other sides are rectangles. (Ex: a pup tent.)

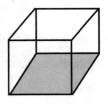

Right Prism: A prism in which the lateral edges are perpendicular to the bases. (Right prisms are assumed unless the problem says otherwise or a diagram is given.)

Oblique Prism: The lateral edges do not form right angles with the bases.

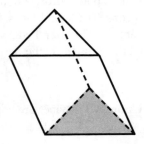

Geometry Made Easy

Two prisms have equal volumes if their bases have equal areas and their altitudes are equal:

Symbols: The area of the base is usually indicated by an upper case B. The altitude (or height) is h. Caution is needed when triangles are involved so the altitude of the triangle and the altitude of the prism are not confused.

Volume of Prisms: The volume of a prism is the area of the base, B, multiplied by the height or altitude of the prism, h. Be careful to make sure that the measurement given for the altitude is the perpendicular distance between the parallel bases. In right prisms it is the same as the length of a lateral edge of the prism. In oblique prisms the altitude is not the length of a lateral edge. Additional work may be required. It doesn't matter what the shape of the prism is, the volume is always the area of the base times the perpendicular distance between the bases.

Examples

❶ The triangular base of the prism in this figure has a base of 12 and an altitude of 7. The height or altitude of the prism is 10. What is its volume in cubic units?

$\text{Area}_{\text{base}} = \dfrac{1}{2}(bh)$ for the triangular base.

$\text{Area}_{\text{base}} = \dfrac{1}{2}(12)(7) = 42$

$\text{Volume} = Bh$
$\text{Volume} = (42)(10) = 420$

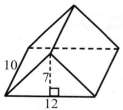

❷ A rectangular prism has the same volume as the triangular prism in the above example (volume = 420). If its base is a rectangle that is 3 by 14, what is its altitude?

$\text{Area}_{\text{base}} = lw$
$\text{Area}_{\text{base}} = (3)(14) = 42$

Volume of triangular prism is 420.
$\text{Volume} = Bh$
$\quad 420 = 42h$
$\quad\quad h = 10$

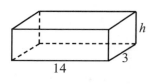

In these examples, even though the shapes are different, the areas of the bases are equal and the altitudes are equal. Therefore, the volume of the two prisms is equal.

Lateral Area: The sum of the areas of the lateral sides only, not including the bases. Use the appropriate formula depending on the shape of the side, rectangle, or parallelogram.

Surface Area: The sum of the areas of each of the faces and the bases equals the surface area of a prism. Find each area separately and add them.

- **Pyramid:** A polyhedron with a regular polygon for a base. The lateral sides are triangles and they meet at a point called a vertex or apex.

Regular Pyramid: The base is a regular polygon (congruent sides, congruent angles). Its lateral sides or faces are congruent isosceles triangles and its lateral edges are congruent. When the altitude, h, is dropped from the apex (or vertex), it meets the base at its center. The

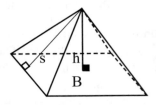

slant height, s, of a regular pyramid is the altitude of one of the congruent isosceles triangular sides. The base can be any type of regular polygon.

Volume: One-third of the product of the area of the base and the altitude (height) of the pyramid. Be careful not to use the slant height in this formula. It may be necessary to use the Pythagorean Theorem or right triangle trigonometry to find the altitude of the pyramid if it is not given.

Example Find the volume of a regular pyramid whose base is a square that measures 10 by 10. The slant height of the pyramid is 13. To solve we must first find the altitude of the pyramid. Since the altitude meets the base in its center, the altitude, the slant height, and a line drawn halfway across the base makes a right triangle. The slant height, 13, is the hypotenuse and the line halfway across the base, 5, is one leg. The altitude is the other leg. This is a 5-12-13 Pythagorean triple – or use the Pythagorean Theorem to find the altitude, h, is 12. The base is a square so its area is product of its sides.

$$V = \frac{1}{3} Bh$$

$$V = \frac{1}{3} (10)(10)(12)$$

$$V = 40$$

Figure 1

Lateral Area: The sum of the triangular sides – not including the base. Use the formula for the area of a triangle for this. Then add. Use figure 1.

Area of one face $\quad A = \frac{1}{2} (10)(13)$

$$A = 75$$

Then multiple by 4. Lateral area of the pyramid is 300.

Surface Area: The sum of all the sides – lateral and base.

Geometry Made Easy

- **Cylinder and Cone:** These figures have circular bases and curved lateral surfaces. Although the bases are circles instead of polygons, cylinders and cones have similar formulas for areas and volume.

Cylinder: Has two congruent circular bases that are parallel. The lateral surface is a curve that connects the circumference of one base to the circumference of the other. Although other types of cylinders do exist, our use of cylinder refers to a right cylinder. Think of a can of soup.

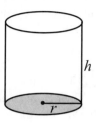

Volume: Multiply the area of one circular base by the altitude (height).

> **Example** What is the volume, in cubic inches, of a can of coffee that has a diameter of 10 inches and is 8 inches tall. Remember to use the radius of 5 in the circle area formula.

Lateral Area: The area of the curved surface of the cylinder. Think about taking the label off a can of soup and laying it out flat. The bottom edge of the label went around the circumference of the base of the soup can, the side of the label is the height of the cylinder.

> **Example** Find the lateral area in square inches of the can of coffee in the above example. Multiply the circumference of the base by the height.

$C_{circle} = 2\pi r$ or πd

$C = 10\pi$

$Lateral\ Area = Ch$

$L.A. = 10\pi(8)$

$L.A. = 80\pi$

Surface Area: Find the lateral area and find the area of the circular bases. Add them.

Cone: Has a circular base and a curved surface for its "sides" which meets in a single vertex opposite the base. Again, we will work with a right cone – one in which the altitude, h, of the cone meets the circular base at its center. With a cone we again sometimes need the *slant height*, s, which is the length of a segment drawn from the vertex to the perimeter of the base.

Volume: $V = (\frac{1}{3})\, Bh$

Example The radius of the base of a cone is 9. The height or altitude of the cone is 10. What is the volume of the cone?

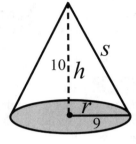

$A_{circular\ base} = \pi r^2$

$A = (9)(9)\pi$

$V = \frac{1}{3}(81\pi)(10)$

$V = 270\pi$

Lateral Area of Cone:

L.A. $= \pi rs$ where s is the slant height of the cone

Example In the figure above find the lateral area of the cone. Round it to the nearest 100th. First we must find s, the slant height. Use the Pythagorean Theorem. The radius is one leg and the height of the cone is the other leg. The hypotenuse is the slant height, s.

$a^2 + b^2 = c^2$

$(10)^2 + (9)^2 = s^2$

$181 = s^2$

$s = \sqrt{181}$ Do not round, leave in radical form.

Now substitute that answer (in radical form) in the lateral area formula.

$L.A. = \pi rs$

$L.A. = (9)(\sqrt{181})$

$L.A. = 380.3922582$ (use your calculator)

$L.A. \approx 380.39$

Surface Area: Add the area of the base and the lateral area. To find it to the nearest 100th, use your calculator and round at the end:

$$2(81\pi) + (9\pi\sqrt{181}) = 634.8612632$$

$$S.\,A. \approx 634.86$$

- **Sphere:** A three dimensional curved figure made up of all the points equidistant from the center. A ball is a sphere.

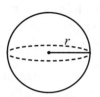

Volume: $V = \dfrac{4}{3}\pi r^3$

Examples

❶ Find the radius of a sphere with a volume of 288π cubic units?

Plug the information known, the volume of 288π, into the formula and solve for r.

$$288\pi = \dfrac{4}{3}\pi r^3$$

$$r^3 = \dfrac{(3)(288\pi)}{4\pi}$$

$$r^3 = 216$$

$$r = 6$$

Surface Area: The surface area of a sphere is the area of the curved surface. The formula is $L.A. = 4\pi r^2$.

❷ Use the answer from example 1 to find the surface area of the sphere in square units.

$$L.A. = 4\pi r^2$$

$$L.A. = 4\pi(6^2)$$

$$L.A. = 144\pi$$

2 — REASONING

Mathematics involves the drawing of conclusions based on parts of mathematics that are interconnected. Logical reasoning skills enable us to draw appropriate conclusions based on the information that is available.

Logic or mathematical reasoning is the analysis of the truth or falsity of a statement based on the Laws of Logic.

Truth Value: Whether a statement (simple or compound) is True or False.

Logically Equivalent: Two or more logic sentences that have the same truth values.

Examples

❶ $(3)(4) = 12$ is logically equivalent to $(2)(6) = 12$
Both statements are true.

❷ $3x = 15$ is logically equivalent to $x = 5$.
Both statements are true.

❸ $4 + 2 = 7$ (*false*) and $4 = 7 - 2$ (*false*) are logically equivalent.

Simple Sentences

Closed Sentence: A closed sentence is one which allows the reader to determine its truth or falsity based on the actual information in the sentence.

Examples

❶ The grass is green.
True – it is normal for grass to be green.

❷ The trees grow flat.
False – Trees do not grow flat, they grow tall.

❸ $3 + 7 = 10$
True – Mathematically we know that three plus seven is ten.

Open Sentence: An open sentence contains a variable (either a word or a mathematical symbol). The truth of an open sentence cannot be determined until the variable is defined.

❶ The house is blue. *True* or *False*? We don't know what house they are talking about so we have no way to know if the sentence is true or not. "House" is the variable here.

❷ $6 + x = 15$ *True* or *False*? We don't know what the value of x is, so we can't tell if that number added to 6 equals 15 or not. "x" is the variable.

Replacement set: A set or group of items that are available to be used to define the variable. Also called the domain.

Solution Set: The members of the replacement set that make the sentence true when they are substituted in place of the variable.

Example If the <u>replacement set</u> or <u>domain</u> is the set of whole numbers, then

❶ The <u>solution set</u> to $6 + x = 15$ is 9.

❷ The <u>solution set</u> to $x > 10$ is {11,12,13, ...}
(**...** means to continue the pattern)

❸ There is no solution to $\frac{2x}{3} = 5$ because the answer, 7.5, is not a whole number and therefore not in the domain.

Negation: The negation of a sentence means that it is the opposite of the sentence given. It also refers to the use of the word "not". The negation of a sentence has the opposite truth value of the original sentence. The negation of a statement is sometimes called the complement of that statement.

Examples

❶ Original sentence: "The house is blue." Its negation is "The house is not blue."

❷ Original sentence: $x + 3 = 10$. Its negation is $x + 3 \neq 10$.

❸ Original sentence: $x < 8$. Its negation is $x \geq 8$

Compound sentences: The truth value of the statement is based on the truth values of its individual phrases. The phrases in the compound sentence are connected with the words, *"and"*, *"or"*, *"if, then"*, and *"if and only if"* or words with the same meaning. Rules are applied to each type of compound statement to determine if the statement is true or false based on given information.

Connectives And Truth Values of Compound Statements

Math Word	Meaning	Rule
Conjunction	AND	*True* only when both phrases are true.
Disjunction	OR	*False* only when both phrases are false.
Conditional	IF, THEN (implies)	*False* only when true implies false.
Biconditional	IF AND ONLY IF	*True* when both parts are true *or* when both are false.

Examples Given the following phrases and the truth value of each.

❶ The sky is blue. *True*

❷ The trees are flat. *False*

❸ The grass is green. *True*

❹ The clouds are red. *False*

Conjunction: "AND" Both parts, called conjuncts, must be True for the statement to be True.

1. The sky is blue *and* the grass is green.
   True — This is a true compound statement because both phrases in it are true.

2. The trees are flat *and* the sky is blue.
   False — One phrase is false but the other is true, based on the given information.

3. The trees are flat and the clouds are red.
   False — Both phrases are false.

Disjunction: "OR" One or both phrases, called disjuncts, must be true for the statement to be true.
1. The sky is blue *or* the grass is green.
    True — Both phrases are true.

2. The trees are flat *or* the grass is green.
    True — One phrase is true.

3. The clouds are red *or* the trees are flat.
    False — Both phrases are false.

Conditional: The statement is written using the words "If, then", "implies", or other wording so the phrases can be rearranged to be interpreted using those words. The conditional is sometimes called an implication.

Hypothesis: The phrase that forms the "if" part of the conditional statement. It does not have to be first in the sentence, but must be the "if" phrase. It is also called the <u>antecedent</u> or <u>premise</u>.

Conclusion: The phrase that makes the "then" part of the conditional statement. Again, it doesn't have to be the last phrase, but must be the "then" phrase. The conclusion is also called the <u>consequent</u>.

THE CONDITIONAL STATEMENT IS *ONLY FALSE* WHEN A TRUE HYPOTHESIS IS USED WITH A FALSE CONCLUSION.

- If the grass is green, then the sky is blue.
    True — Premise is true, consequent is also true. Both phrases are true.

- If the grass is green, then the clouds are red.
    False — The hypothesis is true but the conclusion is false.

- If the trees are flat, then the grass is green.
    True — The hypothesis is false, the conclusion is true.

- If the trees are flat, then the clouds are red.
    True — Both phrases are false. However, the overall sentence is true.

- The clouds are red if the grass is green.
    False — The hypothesis is " the grass is green" which is true and the conclusion is "the clouds are red" which is false. This appears to be "out of order" but can easily be rewritten as "If the grass is green, then the clouds are red."

The truth values of the conditional and its contrapositive are always the same. The inverse and converse may have varying truth values, depending on the content of the hypothesis and conclusion.

Related forms of the Conditional Sentence: There are related forms of writing conditional sentences which may change the meaning of the conditional. The truth value of these sentences can be tracked back to the truth value of the original conditional statement.

Conditional: (original form): hypothesis followed by the conclusion.

Example *If the trees are flat, then the grass is green.*
The hyhpothesis is "the trees are flat".
The conclusion is "the grass is green".

Converse: The phrases in the hypothesis and the conclusion of the conditional change places. *If the grass is green, then the trees are flat.*

Inverse: The phrases in the hypothesis and conclusion of the conditional are negated. *If the trees are not flat, then the grass is not green.*

Contrapositive: The converse and inverse are both performed. The phrases change places and both are negated.
If the grass is not green, then the trees are not flat.

Using the example above: ***If the trees are flat, then the grass is green.***

Conditional: In this particular example the conditional is *true* since the hypothesis is false and the conclusion is true.

Converse: The hypothesis, which has become "the grass is green" is true but the conclusion, "the trees are flat" is false, making the statement *false* overall.

Inverse: The hypothesis, which is now "the trees are not flat" is true and the conclusion, "the grass is not green" is false making the statement *false* overall.

Contrapositive: The hypothesis is now "the grass is not green" which is false and the conclusion is now "the trees are not flat" which is true. Therefore the overall statement is *true*.

In this example, the conditional and contrapositive are true, the converse and inverse are both false.

Examples Here are two math examples of conditionals and their related statements.

❶ **Example of all four statements are true.**

Given: $x = 3$ and $y = 5$ (both are accepted as being true.)

Conditional: If $x + 2 = y$, then $y + 2 = 7$ *True* — Both are True

Converse: If $y + 2 = 7$, then $x + 2 = y$ *True* — Both are True

Inverse: If $x + 2 \neq y$, then $y + 2 \neq 7$ *True* — Both are False

Contrapositive: If $y + 2 \neq 7$, then $x + 2 \neq y$ *True* — Both are False

❷ **Example when conditional and contrapositive are both true and the converse and inverse are both false.**

Given: Every square is a 4 sided polygon which contains 4 right angles.

Conditional: If polygon $ABCD$ is a square, then it contains 4 right angles. *True* —*The truth value of this conditional is true based on the information given (or based on prior knowledge about squares).*

Converse: If polygon $ABCD$ contains 4 right angles, then it is a square. *False* — (It could be a rectangle - we know nothing about the lengths of the sides which must be congruent if it is a square.)

Inverse: If polygon $ABCD$ is not a square, then it does not contain 4 right angles. *False* — (Again - it could be just a rectangle).

Contrapositive: If polygon $ABCD$ does not contain 4 right angles, then it is not a square. *True* — (ALL squares have 4 right angles so if it doesn't have right angles then it can't be a square.)

Logical Equivalence: Two simple or compound sentences are logically equivalent if they always have the same truth value when compared with each other.

The Conditional and its Contrapositive are always logically equivalent.

Biconditional Statements

A biconditional statement is the conjunction of two conditionals. It often involves the words "if and only if". A biconditional is only true when the phrases in it are both true *or* both false. Let's go back to a previous set of simple sentences to demonstrate. (The letters IFF are sometimes used when discussing a biconditional statement.)

Examples Given the following phrases and the truth value of each.

The sky is blue.	*True*
The trees are flat.	*False*
The grass is green.	*True*
The clouds are red.	*False*

❶ The sky is blue if and only if the trees are flat.

Write this as a conjunction of 2 conditionals: "If the sky is blue, then the trees are flat" and "If the trees are flat, then the sky is blue." The first condtional is true, the second is false. Therefore, the biconditional is false.

❷ The sky is blue if and only if the grass is green.

Write this as a conjunction of two conditionals: "If the sky is blue, then the grass is green" and "If the grass is green, then the sky is blue." This biconditional is true, both conditionals are true based on the truth values stated above.

❸ The trees are flat if and only if the clouds are red.

Conditionals: "If the trees are flat, then the clouds are red" and "If the clouds are red, then the trees are flat." Both of these conditionals are True following the rules for conditionals. Therefore, the biconditional is true.

Examples mathematical examples:

❶ **Biconditional:** Triangle *ABC* is equilateral if and only if it is a triangle with 3 congruent sides.

1st Conditional: If triangle *ABC* is equilateral, then it has 3 congruent sides. *True*

2nd Conditional: If a triangle has 3 congruent sides, then it is equilateral. *True*

Since both conditionals are true, the biconditional is *true*.

❷ Biconditional: A quadrilateral is a square if and only if it has 4 congruent sides and 4 right angles.

1ˢᵗ Conditional: If it is a quadrilateral that has 4 congruent sides and 4 angles, then it is a square. *True*

2ⁿᵈ Conditional: If it is a square, then it has 4 congruent sides and 4 right angles. *True.*

∴ Therefore, the biconditional is *true*.

❸ Biconditional: A quadrilateral is a square if and only if it has 4 right angles.

1ˢᵗ Conditional: If it is a quadrilateral with 4 right angles, it is a square. *False* (It could be a rectangle.)

2ⁿᵈ Conditional: If it is a square, then it is a quadrilateral with 4 right angles. *True.*

∴ This biconditional is *false* because the truth values of the conditional parts are not alike.

***Final Note about Logic*:** The reasoning skills used in logic provide the basis for solving many types of math problems and they are especially important in doing geometric proofs.

3 — GEOMETRY
(Proofs and Problem Solving)

Proof: A method of demonstrating the logical sequence of acceptable mathematical reasoning steps used to reach a conclusion. In a problem, certain information is "given" and is accepted as being true. The given information is also called a premise or hypothesis. The solving of the problem, or the "proof" enables us to reach a conclusion - either stated in the problem or determined by analyzing the problem. In order to be certain that the conclusion reached is a correct one, we need to follow logical reasoning steps to get there. We can use hypotheses, axioms, definitions, and proven theorems to demonstrate our thinking process in reaching the conclusion. When doing proofs, it is helpful to note the information you are given on the diagram and also mark each part that you prove as you go along. This gives a better picture of the progress of the proof. At times, it is necessary to change the way you are doing a problem if you see that the steps you have done are not leading to the conclusion desired. Replanning the approach to the problem usually helps.

Note: In this level of math, proofs will relate to figures in one plane. (See also chapter 1 for 3 dimensional geometry information.)

Diagram: A figure that represents the information given in the problem. Labels are very important. Although it doesn't need to be drawn to scale, it is best if the diagram is a fairly good representation of the figure – it will help you think of ideas about solving the problem. As information is developed in the problem, add that information to the diagram – angle or side congruency, perpendicularity, etc. The diagram should be shown as part of the proof on your paper – or as part of the solution to a problem. It can replace a "let" statement in some problems.

Statement: A step that is used in working toward a conclusion in a proof.

Reason: Justification for using a statement. Reasons can be *Postulates - Axioms - Theorems - Corollaries - Properties - Definitions - Formulas*: These are all mathematically acceptable "reasons" which can be used to solve geometry problems and produce geometry proofs. Some texts name things differently - axiom vs. postulate, etc. In this handbook we will usually just call them "reasons". Each student should *classify and use the reasons according to the preferences of his/her teacher and using his/her own textbook.*

Geometry Made Easy

Types of Proofs: Euclidean proofs which are also called statement-reason or formal proofs, paragraph proofs, flow proofs, indirect proofs, and coordinate or analytic proofs. Examples of statement-reason and paragraph proofs are shown throughout the book. Examples of flow proofs and indirect proofs are shown after their descriptions in this section. Coordinate Geometry or analytic proofs are in a separate section – see Ch. 8.

Euclidean Geometry Proof: a formal "statement/reason" or "2-column" proof:
Each step in the progress toward the conclusion is considered to be a "statement" and is written down. Next to the statement the mathematical reason that allowed the step to be done is written. <u>Each step in a proof must be based on steps already completed or on given information</u>. The last step, or statement, will be the conclusion required and next to it, the final "reason" used to get to that conclusion.

EXAMPLES OF EUCLIDEAN PROOFS (Statement/Reason): These problems were chosen to demonstrate the methods used to solve problems using the formal proof process. Check with your teacher for specific instructions, as there are many ways to do a proof.

Examples

❶ Given : $\triangle ABC$
Prove : $m\angle 1 + m\angle 2 + m\angle 3 = 180°$

Statement	Reason
1. $\triangle ABC$	1. Given
2. Through point A, draw line $\overline{NM} \parallel \overline{BC}$ (Label angles as shown.)	2. Through a point not on a given line, there exists one and only one line parallel to the given line
3. $\angle 1 \cong \angle 4,\ \angle 3 \cong \angle 5$	3. If two parallel lines are cut by a transversal, alternate interior angles are $\cong$.
4. $m\angle 1 = m\angle 4;\ m\angle 3 = m\angle 5$	4. Definition of $\cong$ angles.
5. $\angle 4$ and $\angle BAN$ are supplementary.	5. Two angles that form a straight line are supplementary.
6. $m\angle 4 + m\angle BAN = 180°$	6. Definition of supplementary angles.
7. $m\angle BAN = m\angle 2 + m\angle 5$	7. The whole is equal to the sum of the parts.
8. $m\angle 1 + m\angle 2 + m\angle 3 = 180°$	8. Substitution

❷ Given: Lines $\overleftrightarrow{AB}$ and $\overleftrightarrow{CD}$ are cut by transversal $\overleftrightarrow{EF}$

$m\angle 1 = m\angle 5$

Prove: $\overleftrightarrow{AB} \parallel \overleftrightarrow{CD}$

Statement	Reason
1. $m\angle 1 = m\angle 5$	1. Given
2. Lines $\overleftrightarrow{AB}$ and $\overleftrightarrow{CD}$	2. Given
3. $\overleftrightarrow{EF}$ is a transversal	3. Given
4. $m\angle 1 = m\angle 3$	4. Vertical $\angle$'s are equal.
5. $m\angle 3 = m\angle 5$	5. Substitution
6. $\angle 3$ and $\angle 5$ are alternate interior $\angle$'s	6. Definition of alternate interior angles.
7. $\overleftrightarrow{AB} \parallel \overleftrightarrow{CD}$	7. If two lines cut by a transversal form equal alternate interior angles, then the lines are parallel.

❸ Given: Isosceles $\triangle ABC$, $\overline{AC} \cong \overline{AB}$,

$\overline{AD}$ is the $\perp$ bisector of $\overline{BC}$.

Prove: $\triangle ABD \cong \triangle ACD$

Statement	Reason
1. $\overline{AD}$ is the $\perp$ bisector of $\overline{BC}$	1. Given
2. $\triangle ABC$ is isosceles	2. Given
3. $\overline{AC} \cong \overline{AB}$	3. Given
4. $\angle ACB \cong \angle ABD$	4. In an isosceles triangle, the angles opposite the congruent sides are $\cong$.
5. $\overline{CD} \cong \overline{DB}$	5. Definition of $\perp$ bisector
6. $\triangle ABD \cong \triangle ACD$	6. If two sides and the included angle of one triangle are $\cong$ to the corresponding parts of another, the triangles are congruent. SAS $\cong$ SAS

❹ Given: Parallelogram *NEMO*,
diagonal *NCDM*,
$\overline{OC} \perp \overline{NM}$, $\overline{ED} \perp \overline{NM}$
Prove: $\triangle OCM \cong \triangle EDN$

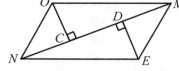

Statement	Reason
1. *NEMO* is a parallelogram $\overline{OC} \perp \overline{NM}$, $\overline{ED} \perp \overline{NM}$	1. Given
2. $\overline{OM} \parallel \overline{NE}$, $\overline{NE} \cong \overline{OM}$	2. Opposite sides of a parallelogram are parallel and congruent.
3. $\angle OCM$ and $\angle EDN$ are right $\angle$'s	3. Perpendicular lines form right $\angle$'s.
4. $\angle OCM \cong \angle EDN$	4. All right $\angle$'s are $\cong$.
5. $\angle NMO \cong \angle MNE$	5. When two parallel lines are cut by a transversal, alternate interior $\angle$'s are $\cong$.
6. $\triangle OCM \cong \triangle EDN$	6. $AAS \cong AAS$

Paragraph Proof: Also called an "informal proof." A plan is made and the statements and reasons are written in the form of a paragraph. Included must be the given information and what is to be proven, a description of the deductive reasoning steps and reasons being used, a diagram when possible, and a conclusion. It could be thought of as writing a formal 2 column proof in a more conversational form, but all the information must be included.

Example Paragraph Proof
Given: Chords $\overline{AB}$ and $\overline{CD}$ of circle *O* intersect at *E*, an interior point of circle *O*; chords $\overline{AD}$ and $\overline{CB}$ are drawn.

Prove: $(AE)(EB) = (CE)(ED)$

We are given chords $\overline{AB}$ and $\overline{CD}$ of circle *O*. They intersect at *E*, an interior point of circle *O*. Chords $\overline{AD}$ and $\overline{BC}$ are drawn. $\angle A \cong \angle C$ because inscribed $\angle$'s of a circle that intercept (subtend) the same arc are $\cong$. $\angle AED \cong \angle CEB$ because they are vertical <'s and are $\cong$. $\triangle AED$ is similar to $\triangle CEB$ because if two $\triangle$'s have two $\angle$'s in one $\cong$ to two $\angle$'s in the other, the triangles are similar. In similar $\triangle$'s, corresponding sides are proportional, so $\dfrac{AE}{CE} = \dfrac{ED}{EB}$. Therefore $(AE)(EB) = (CE)(ED)$ because in a proportion, the product of the means equals the product of the extremes.

Flow Proof: In this type of proof, each statement is written in a box and the reason it it used is written under the box. The boxes are connected with arrows to show the sequence of the proof in reaching the conclusion. Again, the given information, what is to be proven, and a diagram are part of this type of proof.

Examples Flow Proofs

❶ Given: E is the midpoint of $\overline{BC}$ and of $\overline{AD}$

Prove: $\triangle AEB \cong \triangle DEC$

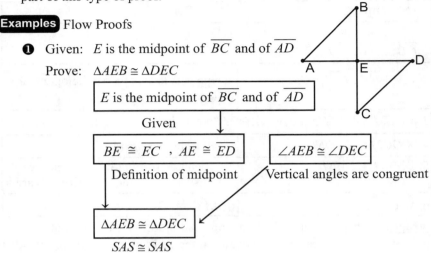

❷ Given: $\angle CAD \cong \angle DAB$
$\overline{AD} \perp \overline{CB}$

Prove: $\overline{AC} \cong \overline{AB}$

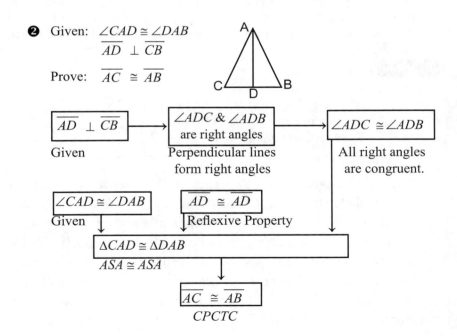

Indirect Proof: Also called "Proof by Contradiction." This proof requires that the we assume the conclusion to be drawn is false! The use of the false conclusion is called an "assumption" and is an important part of this type of proof. Through the progress of the proof, logical reasoning leads to a contradiction of the hypothesis (the "given") or some other known fact such as a theorem, definition, postulate, etc. By reaching an incorrect conclusion based on the assumption (the false original conclusion), we prove the original conclusion must be true.

This is often used to prove that something is not true.

Steps

1) Write the given.

2) Using the "prove" statement, assume it is not true.

3) Proceed as usual trying to prove the assumption in step 2.

4) A contradiction will appear in the proof
-- a statement that is opposite something that is given or known.

5) The contradiction allows the conclusion that the original "prove" statement must be true.

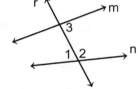

❶ Paragraph Proof By Contradiction:

Given: m is not $\parallel n$

Prove: $\angle 1$ is not $\cong \angle 3$

We are given that line m is not parallel to line n. Assume that $\angle 1 \cong \angle 3$. $\angle 1$ and $\angle 3$ are alternate interior angles by definition. This makes $m\|n$ because if two lines are cut by a transversal and alternate interior angles are congruent, the lines are parallel. This conclusion is contradictory to the given statement. Therefore if $\angle 1$ is not congruent to $\angle 3$, then m is not parallel to n.

❷ Indirect Statement Reason Proof:

Given: $\overline{DE} \cong \overline{DC}$

$\overline{CF}$ is not $\cong \overline{FE}$

Prove: $\angle 1$ is not $\cong \angle 2$

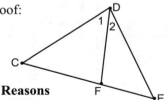

Statement	Reasons
1. $\overline{DE} \cong \overline{DC}$, $\overline{CF}$ is not $\cong \overline{FE}$	1. Given
2. $\angle 1 \cong \angle 2$	2. Assumption
3. $\overline{DF} \cong \overline{DF}$	3. Reflexive Property
4. $\triangle CDF \cong \triangle EDF$	4. $SAS \cong SAS$
5. $\overline{CF} \cong \overline{FE}$	5. CPCTC
6. $\angle 1$ is not $\cong \angle 2$	6. Contradiction step 5 and given.

Geometry Made Easy

Some terms and definitions: This table informally summarizes many of the terms and definitons that are used in plane geometry work. Some of them can be used as reasons in a proof, others are descriptive terms so you can understand terminology in the problem. For more formal definitions, use your text or a math dictionary.

Word, Term, and Mathematical Symbol	Diagram and Labels	Brief Definition
point	• P	A location in space. Has no length, width or depth.
line $\overline{AB}$	A ——— B	Has infinite length, no width.
plane		Has infinite length and width. No depth.
segment $\overline{AB}$	A ——— B	Part of line between 2 points.
collinear points ABC	A B C	Points that are on the same line /line segment.
ray	C D	A line that starts at a point and goes in one direction.
angle $\angle ABC$ or $\angle ABC$	B A C	Formed when 2 rays meet at a point or when 2 lines intersect. Label with the vertex (point) at the center. The size of the opening between the rays is measured in degrees.
bisect line bisector- Figure 1 angle bisector- Fig. 2	Fig. 1 ℓ A B Fig. 2 A B C	To cut in half: A bisector cuts a line segment into 2 $\cong$ parts, an angle into 2 $\cong$ angles.
congruent	B C A E F D	2 or more figures with the same size and shape. Sides are = in length, angles are = in measure.
similar $\triangle MNO \sim \triangle PQR$	M O N P R Q	2 or more figures whose angles have the same measure, corresponding sides are proportional.
perpendicular $\overleftrightarrow{CD} \perp \overleftrightarrow{AB}$	C A B D	2 lines that meet to make a right angle.
parallel $\overleftrightarrow{CD} \parallel \overleftrightarrow{AB}$	A B C D	2 lines in a plane that never meet.
right angle	C B A	Measures 90°.

Geometry Made Easy

straight angle	C	Measures 180°.
acute angle	D	Measures more than 0°, less than 90°.
obtuse angle	R	Measure more than 90°, less than 180°.
reflex angle	S	Measures between 180° and 360°.
equiangular		All angles are congruent.
equilateral		All sides are congruent.
scalene		All sides of a figure are different lengths.
isosceles		2 sides are congruent: (triangles and trapezoids can be isosceles).
regular polygon		A polygon with equal sides & equal angles. 5, 6, 8, 10 and 12 sided polygons are often used.
adjacent angles ∡1 and ∡2 are adjacent ∡s	2 1	Next to each other. Angles which share only one side and a vertex but have no interior points in common.
opposite angles ∡2 and ∡4 are opposite	D 4 3 C A 1 2 B	Across from each other. Not sharing a side or a vertex.
vertex	A B C D C B A	The "point" of an angle, the corner of a polygon.
diagonal	D C A B	Connects 2 opposite vertices (corners) in a geometric figure.
consecutive ∡'s ∡A and ∡B ∡C and ∡D ∡B and ∡C ∡D and ∡A $\overline{AD}$ and $\overline{DC}$ } $\overline{DC}$ and $\overline{CB}$ } consecutive $\overline{BC}$ and $\overline{AB}$ } sides $\overline{AB}$ and $\overline{DA}$	D C A B	Sides or angles that are "one after the other"

	π	
Pi	π is an irrational number. The symbol π should be included in your answer if the problem says leave in terms of Pi .	The ratio of circumference to the diameter of a circle. Use the π button on your calculator to solve a problem involving π.
supplementary angles $m \sphericalangle 1 + m \sphericalangle 2 = 180°$		2 angles whose sum is 180°.
complementary angles $m \sphericalangle 1 + m \sphericalangle 2 = 90°$		2 angles whose sum is 90°.
base $\overline{CD}$ is the base in BCD $\overline{CD}$ and $\overline{AB}$ are both bases in the parallelogram and h is the altitude of each.		In a formula requiring a height measurement, the base of the polygon is the side that the altitude is drawn to. The formula for the area of a trapezoid uses both bases.
altitude and height	$\overline{EF}$ is the altitude in the figure	Perpendicular distance from the base to the opposite vertex is the height. It is the length of the altitude, $\overline{EF}$
Isosceles Δ Vertex angle and base. angles in an isosceles.		ΔABD is isosceles. $\overline{AD} \cong \overline{DB}$ and $\overline{AB}$ is the base. D is the vertex angle, A and B are the base angles.
diameter		Straight line distance from one point on a circle through the center to another point on the circle. (see also ch. #13 Circles)
radius		Distance from the center of a circle to any point on the circle (the edge). (see also ch. #13 Circles)
interior angle of a polygon $C, D, E, \& F$ are all interior $\sphericalangle$'s		An angle formed on the inside of a polygon where the sides intersect.
exterior angle of a polygon $\sphericalangle BCD$ is exterior		An outside angle formed by extending the side of the figure.

Geometry Made Easy

Congruent vs. Equal: Congruent (≅) refers to the shape and size of a geometric figure. Congruent figures have the same shape and have the same measure in degrees (angles) or in length (line segments). Equal (=) refers to the measure of a figure. These terms are used fairly interchangeably in many situations. Even the theorems, which are standard, are written in some texts using the word "congruent" and in others using the word "equal".

When discussing the congruence of line segments, a "hat" is used over the segment. $\overline{AC} \cong \overline{DE}$.
Congruence of angles is shown like this: $\angle ABC \cong \angle DEF$

The length of a segment is shown as: $AC = 10$.
The measure of an angle(s) uses an "m" with the angle symbol:$m\angle A = 75$

- If $AC = 10$ and $DE = 10$, then $\overline{AC} \cong \overline{DE}$.
- If $m\angle A = 75$ and $m\angle B = 75$, then $\angle A \cong \angle B$.

Corresponding Parts: Angles or sides of a polygon that are in matching positions in another polygon (used with similar or congruent polygons).

Properties of Equality: These properties are used as reasons in deductive reasoning geometry proofs. (They are also used in algebra when working with real numbers.)

Addition Property of Equality: If $a = b$, and $c = d$, then $a + c = b + d$
If equal quantities are added to equal quantities, the sums are equal.

Subtraction Property of Equality: If $a = b$, and $c = d$, then $a - c = b - d$.
If equal quantities are subtracted from equal quantities, the differences are equal.

Multiplication Property of Equality: If $a = b$, then $ac = bc$.
If equal quantities are multiplied by equal quantities, the products are equal.

Division Property of Equality: If $a = b$ and $c \neq 0$, then $a \div c = b \div c$.
If equal quantities are divided by equal quantities which are $\neq 0$, the quotients are equal.

Reflexive Property of Equality: Anything is equal to itself.
[number, line segments, angles]

Symmetric Property of Equality: If $a = b$, then $b = a$.

Transitive Property of Equality: If $a = b$ and $b = c$, then $a = c$.

Substitution Property of Equality: If $a = b$, then "a" can be replaced by "b" and "b" can be replaced by "a". [If $a + b = c$ and $m + n = c$, then $a + b = m + n$ using substitution.]

Partition Postulate: The whole is equal to the sum of its parts. This is also called "**Segment Addition**" or "**Angle Addition**" or "**Betweeness**" depending on its use in a problem.

Betweeness: If point B is between point A and point C, then $\overline{AB} + \overline{BC} = \overline{AC}$

Very commonly used definitions:

Distance:
1) The length of a segment between two points.
2) The length of the perpendicular segment from a point to a line. Also called the perpendicular distance. This is the "default" – always use the perpendicular distance unless directed otherwise.
3) Equidistant means "equally distant." Example: any two or more points on a circle are equidistant from the center.

Midpoint Definition: If point M, on a line segment, $\overline{AC}$, divides the segment into two $\cong$ segments, $\overline{AM} \cong \overline{MC}$, then point M is the midpoint of $\overline{AC}$.

Definition of Bisector:
1) Angle Bisector: An angle bisector divides an angle into two congruent angles.
2) Bisector of a segment: The bisector of a line segment divides a line segment into two congruent line segments.

Points and Lines in a Plane: (See chapter 1 for more detailed 3 dimensional information.)
- Through any two points there is exactly one straight line.
- If two lines intersect in a plane, their intersection is a point.
- If two points lie in a plane, then the line joining them lies in that plane.
- In a plane, two lines that do not intersect are parallel lines.
- In a plane, two lines that inersect to form right angles are perpendicular.
- Through any three noncollinear points there is exactly one plane.
- Through any point not on a given line, there exists one and only one line parallel to the given line in that plane.
- Through any point not on a given line, there exists one and only one line perpendicular to the given line in that plane.

Note: The information on pages 24 – 27 can be used in writing proofs and solving problems in addition to the more complex information that follows.

When and how to use "reasons" in a proof: In this section we will show how the relationship of theorems, definitions, postulates, corollaries, etc. might be used. The format is "known" which would be the "given" in a problem - or a step that you've already completed. The "show" is what must be proven or what is needed for the next step to be processed in the solution or proof. All of the "reasons" can be used in problem solving as well as in proofs.

ANGLES, PARALLEL LINES, TRANSVERSALS, AND PERPENDICULAR LINES:

1) Known: *Lines are parallel* and are cut by a transversal.
    Show: Angles are congruent or find the measures of the angles.
    (Parallelism is known, find angles. Sometimes used to prove triangles are ≅ within a parallelogram.)

    –   If two lines are parallel and are cut by a transversal, then corresponding angles are equal.

    –   If two parallel lines are cut by a transversal, then alternate interior angles are equal.

    –   If two parallel lines are cut by a transversal, then same-side interior angles
        are supplementary.

    –   If three parallel lines intersect two transversals, they divide the transversals proportionally. $a \parallel b \parallel c$.

    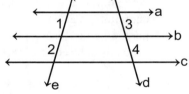

    Segments are proportional: $1 : 2 = 3 : 4$

2) Known: Two lines are cut by a transversal and some of *the angles are* ≅ .
    Show: The lines are parallel.(Angles are known, prove parallelism of lines. Often used to prove a quadrilateral is a parallelogram.)

    –   If two lines and a transversal form equal corresponding angles, then the lines are parallel.

    –   If two lines and a transversal form equal alternate interior angles, then the lines are parallel.

    –   If two lines are cut by a transversal and form supplementary same-side angles, then the lines are parallel.

3) Known: Two lines are perpendicular ($\perp$).
   Show: Angles are right angles, angles = 90°, parallelogram is a rectangle, or a triangle is a right triangle.

   – Perpendicular lines form right angles.

   – All right angles are congruent. (This is frequently used when perpendicular lines are given.)

   – In a plane, if two lines are each perpendicular to a third line, the lines are parallel to each other.

4) Known: Perpendicular bisector of a segment.
   Show: Congruency of segments drawn to it.

   – Any point on the perpendicular bisector of a segment is equidistant from the endpoints of the segment.

   – A $\perp$ bisector divides the segment it bisects into two $\cong$ segments.

   – A $\perp$ bisector forms right angles with the segment it bisects.

5) Known: Segments are congruent, $\cong$.
   Show: Perpendicular bisector.

   – Any point that is equidistant from the endpoints of a segment is on the perpendicular bisector of that segment. Line c is $\perp$ bisector of $\overline{AB}$.

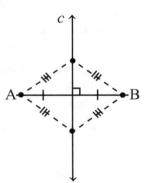

6) Known: Angles are given.
   Show: congruency of angles.

   – Vertical angles are congruent, $\cong$.

   – All right angles are congruent, $\cong$.

   – An angle bisector divides an angle into two $\cong$ angles.

   – If two angles are complements of equal angles (or of the same angle), then the two angles are $\cong$.

   – If two angles are supplements of equal angles (or of the same angle), then the two angles are $\cong$.

   – Corresponding parts of congruent triangles are $\cong$.
     (Must prove or know that the triangles are $\cong$ first.)

Parallel Lines Proof

Given: $n\|m$, $t\|s$

Prove: $\angle 1 \cong \angle 10$

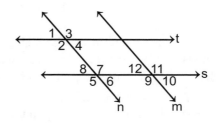

Statement	Reason
1. $n\|m$, $t\|s$	1. Given
2. $\angle 1 \cong \angle 8$	2. If two parallel lines are cut by a transversal, corresponding angles are $\cong$. (*t* and *s* are the parallel's, *n* is the transversal.)
3. $\angle 8 \cong \angle 6$	3. Vertical angles are $\cong$.
4. $\angle 1 \cong \angle 6$	4. Transitive Property (or Substitution)
5. $\angle 6 \cong \angle 10$	5. If two parallel lines are cut by a transversal, corresponding angles are $\cong$. (*n* and *m* are the parallel's, *s* is the transversal.)
6. $\angle 1 \cong \angle 10$	6. Transitive Property (or Substitution.)

4 — POLYGONS

Polygon: A plane figure in which all sides are line segments. A polygon is named by the number of sides it contains. In general, a polygon with "n" sides is called an "n-gon". (Ex: A hexagon can also be called a 6-gon).

We will work primarily with *convex polygons* in which all the all the vertices point outward - think of the shape of a stop sign.

In a *concave* polygon, at least one pair of sides points inward making a cavelike appearance to that side - think of a star shaped polygon.

1) Known: A specific n-gon is described. It is convex.
2) Show: The measures of its **interior** and/or **exterior** angles.

– The formula $S = (n - 2)(180°)$ gives the **interior** angle sum of a convex polygon with n sides.
$\angle 1 + \angle 2 + \angle 3 + \angle 4 + \angle 5$
$= (5 - 2)(180) = 540°$

– The sum of the **exterior** angles of any convex polygon is 360°.
$\angle a + \angle b + \angle c + \angle d + \angle e = 360°$

– An interior angle and its adjacent exterior angle are supplementary angles $\angle 1 + \angle a = 180°$. If information is given that enables you to know the measure of an exterior angle, you can find the interior angle by subtracting from 180 degrees. Use the same process if you know the measure of the interior angle to find the adjacent exterior angle.

Regular Polygon: A polygon in which all sides are congruent *and* all angles are congruent.

1) Known: A regular polygon is either given or described and it has n sides.
2) Show: The value of its interior and/or exterior angles.

– The sum of the exterior angles in any convex polygon is 360°. Since the sides are all in a regular polygon, *360/n* will give the measure of an exterior angle of an n-sided regular polygon.

– The sum of the interior angles of any convex polygon is $(n - 2)(180)$ and since all the angles are equal, to find the measure of one interior angle divide by n.

Angles of Polygon Examples:

Examples

❶ Find the measure of an interior angle of an 8 sided regular polygon (octagon). $n = 8$

Formula: $\dfrac{(n-2)180}{n}$ An interior $\angle$ of a regular octagon measure $135°$

$$\dfrac{(n-2)(180)}{8} = 135°$$

❷ The exterior angle of a regular n-gon measures $24°$.
Find the value of n.
Solution: The sum of the exterior angles of any n-gon $= 360°$.
Divide $360°$ by the measure of one exterior angle in a regular polygon to find the number of sides.

$$\dfrac{360}{24} = n$$

$$n = 15$$

Congruent Polygons: 2 (or more) polygons that have corresponding sides that are congruent and corrresponding angles that are congruent.

Similar Polygons: Polygons which have corresponding angles that are equal and corresponding sides that are proportional.

FORMULAS FOR POLYGONS: Formulas are not often seen as part of a formal geometry proof, but they are mathematically acceptable procedures to be used in problem solving and in coordinate geometry problems. When using the formulas remember what you know about congruency or ratios in certain figures, the Pythagorean Theorem $c^2 = a^2 + b^2$, the distance formula, $d = \sqrt{(x_2 - x_1)^2 + (y_2 - y_1)^2}$, and any other prior knowledge you may have about finding the lengths of segments needed in the formula.

Note: Remember that height (h) means the perpendicular distance from the base to the vertex opposite it. The length of the altitude is the height. Squares and rectangles have specific formulas because the sides of the figure, being already perpendicular, are the base and height of the figure. In figures without a right angle, height is a not a side and is sometimes shown by a dotted line in a drawing.

[Note: Because the height of a figure is perpendicular to the base, it is often possible to use the Pythagorean Theorem to find the height, see page 47.]

SIMILAR POLYGONS AND PROPORTIONS:

- If two polygons are similar, each $\angle$ in one is $\cong$ to the corresponding $\angle$ of the other, and each side in one is proportional to the corresponding side of the other.

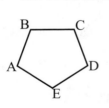 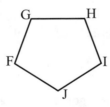

$ABCDE \sim FGHIJ$

$\angle A \cong \angle F \quad \angle D \cong \angle I$

$\angle B \cong \angle G \quad \angle E \cong \angle J$

$\angle C \cong \angle H$

$$\frac{AB}{FG} = \frac{BC}{GH} \text{ etc.}$$

(2 Figures – same shape, different sizes.)

- The scale factor or ratio factor of two similar polygons is the ratio of the side of one to the corresponding side of the other.

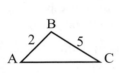 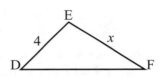

$\triangle ABC \sim \triangle DEF$

$$\frac{2}{4} = \frac{1}{2}$$

$\frac{1}{2}$ is the ratio factor of $\triangle ABC : \triangle DEF$

(Use it to find missing sides.)

$$\therefore \frac{1}{2} = \frac{5}{x} \Rightarrow x = 10$$

- If two polygons are similar and have a scale factor of $a : b$, the ratio of their perimeters is $a : b$.

 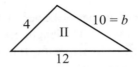

$\triangle I \sim \triangle II$

Ratio of $\dfrac{a}{b} = \dfrac{5}{10} = \dfrac{1}{2}$

Perimeter $\triangle I = 16$

Perimeter $\triangle II = 32$

$$\frac{P_I}{P_{II}} = \frac{16}{32} = \frac{1}{2}$$

- If two polygons are similar and have a scale factor of $a : b$, the ratio of their areas is $a^2 : b^2$.

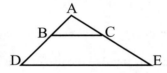

12

$a = 2$ I   4

$b = 6$ II

Rectangle I $\sim$ Rectangle II

Ratio of sides $= \dfrac{a}{b} = \dfrac{2}{6} = \dfrac{1}{3}$

$\left.\begin{array}{l} \text{Area}_\text{I} = 2(4) = 8 \\ \text{Area}_\text{II} = 6(12) = 72 \end{array}\right\}$ $\left(\dfrac{a}{b}\right)^2 = \dfrac{A_\text{I}}{A_\text{II}}$

$\left(\dfrac{1}{3}\right)^2 = \dfrac{8}{72} = \dfrac{1}{9}$

- If a line intersects the sides a triangle and is parallel to the third side, then the line divides the two intersected sides proportionally.

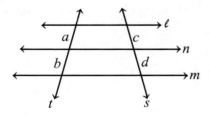

$\overline{BC} \parallel \overline{DE}$

$\dfrac{\overline{AC}}{\overline{CE}} = \dfrac{\overline{AB}}{\overline{BD}}$

- If three parallel lines intersect two transversals, they divide the transversals proportionally.

$\ell \parallel n \parallel m$

t and s are transversals.

$\dfrac{a}{b} = \dfrac{c}{d}$

5 — TRIANGLES

Triangle: A triangle is a polygon with 3 sides.

Angles in a triangle: The sum of the angles in a triangle = 180°.

Types of triangles
- Classified by the measure of their angles
  1. Acute: Each of the 3 angles is less than 90°
  2. Right: One angle is 90°, the other two are a complementary pair.
  3. Obtuse: One angle is greater than 90° and less than 180°
  4. *Equiangular: All 3 angles are congruent (60° each)

- Classified by the lengths of their sides
  1. Scalene: none of the 3 sides are congruent
  2. Isosceles: 2 sides are congruent
  3. *Equilateral: all 3 sides are congruent.

  *Equilateral triangles are also equiangular. They are regular polygons.

Many different types of geometry problems involve the use of triangles. We are asked to use them in situations where they are congruent, similar, parts of larger problems involving other polygons, and circles. Many other applications are possible.

There are three methods used to prove that triangles are similar. See page 40.

There are four methods used to prove that triangles are congruent and one for right triangles only. See page 37.

TRIANGLE CONGRUENCIES: "Included" means it is between the other two parts described.

1) Show: Triangles are congruent. Use any of the following reasons.

- ASA ≅ ASA. (angle-side-angle)
  If two angles and the included
  side in one triangle are congruent
  to the corresponding parts of the
  other, the triangles are ≅.

- SAS ≅ SAS. (side-angle-side)
  If two sides and the included
  angle in one triangle are ≅ to the
  corresponding parts of the other,
  the triangles are ≅.

- SSS ≅ SSS. (side-side-side)
  If three sides of one triangle
  are ≅ to three sides of another,
  the two triangles are congruent.

- AAS ≅ AAS. (angle-angle-side)
  If two angles and a non-included
  side of one triangle are ≅ to the
  corresponding parts of another,
  the triangles are ≅.

- HL Hypotenuse - (Leg Theorem)
  If the hypotenuse and one leg
  of a right triangle are ≅ to the
  corresponding parts of the other,
  the two triangles are ≅.Triangles
  must be identified as right triangles
  before using HL.

Note: SSA ≅ SSA is **not** an adequate proof for triangle congruency due to the "Ambiguous Case" which will be studied in Trigonometry.

❶ Given: $\overline{BE}$ intersects $\overline{AD}$ at C.
$\angle B \cong \angle D, \overline{BC} \cong \overline{DC}$

Prove: $\triangle ABC \cong \triangle EDC$

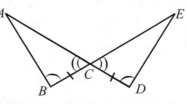

Statement	Reason
1. $\angle B \cong \angle D, \overline{BC} \cong \overline{DC}$	1. Given
2. $\angle ACB \cong \angle ECD$	2. Vertical angles are $\cong$.
3. $\triangle ABC \cong \triangle EDC$	3. $ASA \cong ASA$

❷ Given: $\overline{ED} \cong \overline{AC}$ bisect each other.

Prove: $\triangle ABE \cong \triangle CBD$

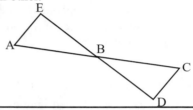

Statement	Reason
1. $\overline{ED} \cong \overline{AC}$ bisect each other.	1. Given
2. $\overline{AB} \cong \overline{CB}, \overline{EB} \cong \overline{DB}$	2. Definition of bisector
3. $\angle ABE \cong \angle CBD$	3. Vertical angles are $\cong$.
4. $\triangle ABE \cong \triangle CBD$	4. $SAS \cong SAS$

❸ Given: Rectangle $ABCD$.
Diagonal $\overline{BD}$
Prove: $\triangle ABD \cong \triangle CBD$

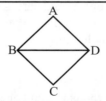

Statement	Reason
1. $ABCD$ is a rectangle. $\overline{BD}$ is a diagonal.	1. Given
2. $\overline{AD} \cong \overline{BC}$ $\overline{AB} \cong \overline{CD}$	2. In a rectangle opposite sides are $\cong$.
3. $\overline{BD} \cong \overline{BD}$	3. Reflexive Property
4. $\triangle ABD \cong \triangle CBD$	4. $SSS \cong SSS$

❹ Given: $\angle CAE \cong \angle CBD$
$\overline{CD} \cong \overline{CE}$
Prove: $\triangle CDB \cong \triangle CEA$

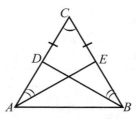

Paragraph Proof:

$\angle CAE \cong \angle CBD$ and $\overline{CD} \cong \overline{CE}$ because this is the given information. $\angle DCE \cong \angle ECD$ by the reflexive property. The two triangles, $\triangle CDB$ and $\triangle CEA$ are congruent because one triangle has two angles and one side congruent to the corresponding parts of the other.

Therefore: $\triangle CDB \cong \triangle CEA$

Note: When overlapping triangles are involved, it often helps to use a colored pencil and outline the two triangles you are working with different colors in order to see "what's what" more easily!

---

❺ Given: $\overline{AC} \perp \overline{DB}$
$\overline{AD} \cong \overline{AB}$
Prove: $\triangle DCA \cong \triangle BCA$

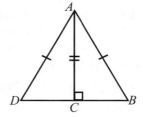

Statement	Reason
1. $\overline{AC} \perp \overline{DB}$, $\overline{AD} \cong \overline{AB}$	1. Given
2. $\triangle ACD$ and $\triangle ACB$ are right angles.	2. $\perp$ lines form right angles.
*3. $\triangle DCA$ and $\triangle BCA$ are right triangles.	3. Definition of right triangle.
4. $\overline{AC} \cong \overline{AC}$	4. Reflexive Property
5. $\triangle DCA \cong \triangle BCA$	5. $HL \cong HL$

* The triangles must be proven and stated to be to be right triangles before using $HL \cong HL$.

---

❻ Given: $\overline{AB} \cong \overline{AE}$
$\overline{BC} \cong \overline{DE}$

Prove: $\triangle ACD$ is isosceles

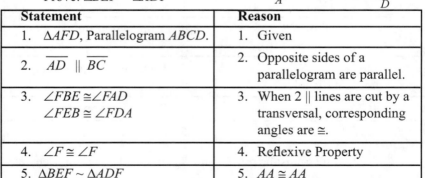

Statement	Reason
1. $\overline{AB} \cong \overline{AE}$, $\overline{BC} \cong \overline{DE}$	1. Given
2. $\angle B \cong \angle E$	2. In $\triangle ABE$, the angles opposite $\cong$ sides are $\cong$.
3. $\triangle ABC \cong AED$	3. $SAS \cong SAS$
4. $\overline{AC} \cong \overline{AD}$	4. CPCTC
5. $\triangle ACD$ is isosceles	5. A triangle with two $\cong$ sides is isosceles.

TRIANGLE SIMILARITY: In proving that triangles are similar, we can use three different methods.

1. AA: Prove that two of the three angles in one triangle are congruent to two angles in the other.

2. SSS: Prove that each side of one triangle is proportionate to the corresponding side of the other triangle.

3. SAS: Prove that two sides of one triangle are proportionate to the two corresponding sides of the other and that the included angle is congruent in both triangles.

SEE ALSO: Right triangles page 48, Dilations, page 84

Examples **Triangle Similarity Proofs**

❶ Statement Reason Proof
Given: $\triangle AFD$. Parallelogram $ABCD$.
Point B is on side $\overline{AF}$ and E is on $\overline{BC}$.
Prove: $\triangle BEF \sim \triangle ADF$

Statement	Reason
1. $\triangle AFD$, Parallelogram $ABCD$.	1. Given
2. $\overline{AD} \parallel \overline{BC}$	2. Opposite sides of a parallelogram are parallel.
3. $\angle FBE \cong \angle FAD$ $\angle FEB \cong \angle FDA$	3. When 2 $\parallel$ lines are cut by a transversal, corresponding angles are $\cong$.
4. $\angle F \cong \angle F$	4. Reflexive Property
5. $\triangle BEF \sim \triangle ADF$	5. $AA \cong AA$

(These 2 proofs use the same diagram, but different triangles are to be proven similar.)

Geometry Made Easy

❷ Paragraph Proof

Given: Parallelogram *ABCD*.
$\overline{AB}$ is extended to *F* and $\overline{FD}$ is
drawn. $\overline{FD}$ intersects $\overline{BC}$ at *E*.

Prove: $\triangle BEF \sim \triangle CED$

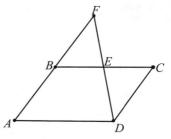

We are given parallelogram *ABCD* which makes $\overline{AB} \parallel \overline{DC}$ and
$\overline{BC} \parallel \overline{AD}$ by definition. $\angle FBE \cong \angle DCE$ and $\angle BFE \cong \angle CDE$ because
when parallel lines are cut by a transversal, alternate interior angles are
congruent. $\angle FEB \cong \angle DEC$ because vertical angles are congruent.
Therefore $\triangle BEF \sim \triangle CED$ because when two angles in a triangle are
congruent to two angles in another triangle, the triangles are similar.

❸ Given: $\triangle ABC$. *E* is the midpoint of $\overline{BC}$,
D is the midpoint of $\overline{AB}$.

Prove: $\triangle ABC \sim \triangle EBD$

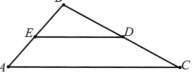

Statement	Reasons
1. $\triangle ABC$, *D* is the midpoint of $\overline{BC}$, *E* is the midpoint of $\overline{AD}$	1. Given
2. $\overline{ED} \parallel \overline{AC}$ and $\frac{ED}{AC} = \frac{1}{2}$	2. A segment joining the midpoints of 2 sides of a triangle is parallel to the 3rd side and equal to $\frac{1}{2}$ of its measure.
3. $BD = \frac{1}{2}(BC)$, $BE = \frac{1}{2}(AB)$	3. A midpoint divides a segment in half.
4. $\triangle ABC \sim \triangle EBD$	4. Two triangles are similar if 3 sides of one are proportional to the corresponding sides of he other.

Note: This proof could be done using *AA* as well. Since $\angle B$ is shared
by both triangles, it equals itself by the reflexive property. Since
$\overline{ED} \parallel \overline{AC}$, $\angle BED \cong \angle BAC$ and $\angle BDE \cong \angle BCA$ because they are
pairs of corresponding angles which are congruent. That makes
the two triangles similar by *AA*.

Using calculations in Similar Triangles

❶ Two triangles are similar. The sides of one triangle measure 23, 27, and 40. The middle side of the second triangle is 81. Find the lengths of the shortest side and longest side of the second triangle.

Solution: Find the ratio of the measures of the middle sides – they are the corresponding sides whose measures are given in the two triangles.

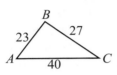
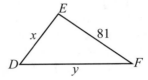

$$\triangle ABC \sim \triangle DEF$$

$\dfrac{27}{81} = \dfrac{1}{3}$ Use this ratio to find the values of the other sides by making proportions.

$$\frac{23}{x} = \frac{1}{3} \text{ and } \frac{40}{y} = \frac{1}{3}$$
$$x = 69 \qquad y = 120$$

The shortest side of the second triangle is 69 and the longest side is 120.

❷ $\triangle ABC \sim \triangle DEF \sim \triangle GHI$

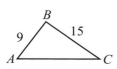 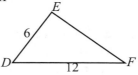

The ratio factors between all three triangles are not always the same. Choose two triangles to work on at a time to find the lengths of the missing sides.

a) Use $\triangle ABC$ and
$\triangle DEF$ to find AC.

$$\frac{AB}{DE} = \frac{AC}{DF}$$

$$\frac{9}{6} = \frac{AC}{12}$$

$$6(AC) = 9(12)$$

$$\boxed{AC = 18}$$

b) Use $\triangle ABC$ and $\triangle DEF$
again to find EF.

$$\frac{AB}{DE} = \frac{BC}{EF}$$

$$\frac{9}{6} = \frac{15}{EF}$$

$$9(EF) = 6(15)$$

$$\boxed{EF = 10}$$

c) Use either $\triangle ABC$ and $\triangle DEF$
with $\triangle GHI$ to find GH.

$$\frac{DE}{GH} = \frac{DF}{GI}$$

$$\frac{6}{GH} = \frac{12}{6}$$

$$6(6) = 12(GH)$$

$$\boxed{GH = 3}$$

General Triangle Information
Triangles and Angles

- The sum of the angles in any triangle is 180°.
  $m\angle 1 + m\angle 2 + m\angle 3 = 180$

- An exterior angle of a triangle is equal to the sum of the two opposite (remote, nonadjacent) interior angles. $\angle 4 = \angle 2 + \angle 3$

- An exterior angle of a triangle is greater than either remote interior angle. $\angle 4 > \angle 2, \angle 4 > \angle 3$

- If two angles in a triangle are equal to the corresponding angles in another triangle, then the third angles in the triangles are ≅. This does not prove congruency of the triangles by itself. It must be combined with information about at least one pair of corresponding sides. It can be used to prove similarity of triangles.

Triangles and Sides

- The sum of any two sides of a triangle must be greater than the third side.

- Midsegments or Midlines: The segment joining the midpoints of two sides of a triangle is parallel to the third side and half as long as the third side.

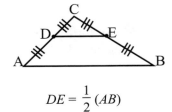

$$DE = \frac{1}{2}(AB)$$

- Base of a triangle
  1. Any triangle: When an altitude of a triangle is drawn, the side it is perpendicular to is called the "base" of the triangle.
  2. Isosceles triangle: The base of an isosceles triangle is the side that is not one of the two congruent sides.
  3. Equilateral triangle: Any of the three sides can be considered the base. If an altitude is drawn, the side perpendicular to it is the base.

Relationships with sides and angles of a triangle:

- If two sides of a triangle are ≅, then the angles opposite those sides are equal. (isosceles triangle)

- If two angles of a triangle are ≅, the sides opposite them are ≅. (isosceles)

- Base angles of an isosceles triangle are ≅. (Remember base angles are opposite the congruent sides.)

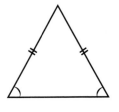

- If a triangle is equiangular (equal angles), it is also equilateral (equal sides).

- An equilateral triangle is also equiangular and each angle measures 60°.

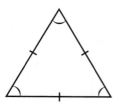

- In a triangle, the side opposite the largest angle is the largest side. (The other two sides have the same relationship to their opposite angles. The middle sized side is opposite the middle sized angle and the smallest side is opposite the smallest angle.)

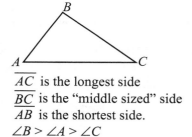

$\overline{AC}$ is the longest side
$\overline{BC}$ is the "middle sized" side
$\overline{AB}$ is the shortest side.
$\angle B > \angle A > \angle C$

- In a right triangle, the hypotenuse, the side opposite the right angle, is always the longest side. $\overline{AB}$ is the longest side and is the hypotenuse.

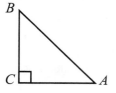

- Corresponding parts of congruent triangles are congruent. (Use for corresponding sides or corresponding angles.) [Often abbreviated as CPCTC.]

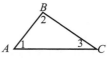 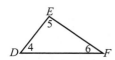

If $\triangle ABC \cong \triangle DEF$, then $\overline{AB} \cong \overline{DE}$, $\overline{BC} \cong \overline{EF}$, $\overline{AC} \cong \overline{DF}$ and
$\angle 1 \cong \angle 4$, $\angle 2 \cong \angle 5$, $\angle 3 \cong \angle 6$

Isosceles Triangles: This information is included with the general triangle information, but it is specific to isosceles triangles.

- Known: A triangle is isosceles (given, or already proven)
  - In an isosceles triangle, the angles opposite the congruent sides (base angles) are congruent.
  - In an isosceles triangle, two sides are congruent.
  - The altitude of an isosceles triangle drawn from the base divides the triangle into two congruent right triangles.

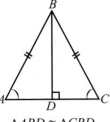

$\triangle ABD \cong \triangle CBD$

Note: The altitude of an isosceles triangle can often be found using the pythagorean theorem. (See page 48)

RIGHT TRIANGLES: [Remember, the *hypotenuse* is the side opposite the right angle and is always the longest side. The *legs* meet to form the right angle.]

Hypotenuse - Leg Theorem: In two right triangles, if the hypotenuse and a leg of one are congruent to the corresponding parts of the other, the triangles are congruent. It must be stated that the triangles are right triangles first.

Additional Right Triangle information:
The midpoint of the hypotenuse of a right triangle is equidistant from the three vertices.

The two acute angles in a right triangle are complementary.

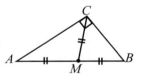

Pythagorean Theorem: "*c*" is the *hypotenuse* and "*a*" and "*b*" are the legs. Side "*c*" is the longest side. If a triangle is a right triangle, then the square of the length of the hypotenuse is equal to the sum of the squares of the lengths of the legs. As a formula it is written: $c^2 = a^2 + b^2$

The Pythagorean Theorem is also commonly seen in its converse form which is, "If the sum of the squares of the lengths of the two legs of a triangle equals the square of the hypotenuse, the triangle is a right triangle. $a^2 + b^2 = c^2$

<u>Note:</u> The Pythagorean Theorem is used to find a missing side of a triangle, and it also can be used to prove a triangle is a right triangle. If the sides of a given triangle can be substituted into the Pythagorean Theorem, using the longest side as "*c*" and the numbers "check", then the triangle is a right triangle.

Geometry Made Easy

❶ Known: Measures of two sides of a right triangle.
Show: Measure of the third side.

The hypotenuse of a right Δ is 17.
One leg is 8. Find the lenght of
the other leg.

$8^2 + b^2 = 17^2$

$64 + b^2 = 289$

$b^2 = 225$

$b = 15$

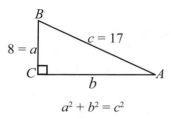

$a^2 + b^2 = c^2$

❷ Known: Measure of three sides of a triangle.
Show: The triangle is or is not a right triangle.

– In a triangle, if the square of the longest side equals the sum of the
squares of the other two sides, the triangle is a right triangle.

Is a triangle with sides of 5, 10, and $5\sqrt{3}$ a right triangle?

Solution: If the sides "check" in the Pythagorean Theorem, it is a
right triangle. The longest side is the hypotenuse. ($5\sqrt{3} \approx 8.66$).

$\left(5\sqrt{3}\right)^2 + (5)^2 = 10^2$

$75 + 25 = 100$

$100 = 100$ √

This is a right triangle.

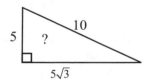

❸ Is a triangle with sides of 6, 10, and 12 a right triangle?

Solution: Only sides that "check" in the Pythagorean Theorem can
form right triangles. The hypotenuse is always the longest side.

$6^2 + 10^2 = 12^2$ Does it work?

$36 + 100 = 144$

$136 \neq 144$

This is **NOT** a right triangle

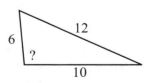

Isosceles Triangles and The Pythagorean Theorem

Since the altitude of an isosceles triangle is the perpendicular bisector of the base of the triangle, the Pythagorean Theorem is often used in calculations for isosceles (or equilateral) triangles.

Example

Find the height of isosceles triangle ABC. The base, $\overline{AC}$, is 10 inches and the congruent sides are each 14 inches.

Solution: The altitude divides the base into 2 equal segments, each 5 inches. The 14 inch side becomes the hypotenuse formed when the altitude is drawn.

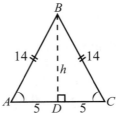

$a^2 + b^2 = c^2$
$5^2 = b^2 = 14^2$
$b^2 = 196 - 25$
$b = \sqrt{171}$ inches
∴ The height, h,
is $\sqrt{171}$ inches

Pythagorean Triples: There are some common right triangle measurements or mutliples of them that are helpful to recognize. They are called Pythagorean Triples. Examples: 3, 4 and 5; 5, 12, 13; and 8, 15, 17. Remember the longest side is always the hypotenuse. There are many more Pythagorean Triples but these are the most common.

Proportions in Right Triangles

Altitude Drawn to the Hypotenuse of a Right Triangle:

1) Similar triangles are formed

When the altitude of a right triangle is drawn to the hypotenuse, two similar triangles are created. There are some theorems that can be applied directly to this situation, but my students have found it helpful to see the similarity of the triangles as well.

In triangle ABC, C is the right angle. The altitude, $\overline{CD}$ is drawn to hypotenuse $\overline{AB}$. The diagram shows the relationship between the three triangles – they are taken apart to show them more clearly. The similarity relationship is true no matter what the acute angle measures are in the original right triangle. The example on the next page has angles of 90°, 25°, and 55° to demonstrate.

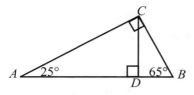

$\triangle ABC$	$\triangle ADC$	$\triangle CDB$
$\sphericalangle C = 90°$	$\sphericalangle D = 90°$	$\sphericalangle D = 90°$
$\sphericalangle A = 25°$	$\sphericalangle A = 25°$	$\sphericalangle DCB = 25°$
$\sphericalangle B = 65°$	$\sphericalangle DCA = 65°$	$\sphericalangle B = 65°$

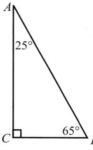
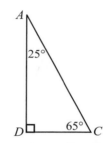

Since all three triangles have corresponding angles that are congruent, the triangles are all similar.

2) Mean Proportionals in a Right triangles with an altitude:

Known: An altitude is drawn to the hypotenuse of a right triangle

Show: Relationships between segments of a right triangle.
Find length of segments.

- The altitude drawn to the hypotenuse of a right triangle divides the triangle into two triangles which are similar (~) to each other and similar to the original triangle. $\triangle ADC \sim \triangle CDB \sim \triangle ACB$.

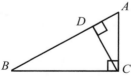

- In a right triangle with the altitude drawn to the hypotenuse, the length of each leg is the mean proportional between the length of the segment of the hypotenuse that is adjacent to the leg and the whole length of the hypotenuse. $\dfrac{AD}{AC} = \dfrac{AC}{AB}$ and $\dfrac{BD}{BC} = \dfrac{BC}{AB}$

Note: The part of the hypotenuse that is adjacent a leg is sometimes called, "The shadow of the leg." In the example below $\overline{BD}$ is the shadow of $\overline{BC}$ and $\overline{AD}$ is the shadow of $\overline{AC}$.

Right triangles with an altitude

❶ In $\triangle ABC$, CD is the altitude drawn to hypotenuse $\overline{AB}$.
$BC = 10$, $BD = 4$. Find the length of AB.

Solution: Leg $\overline{BC}$ is the mean proportional between the segment
of the hypotenuse it is adjacent to and the entire hypotenuse.

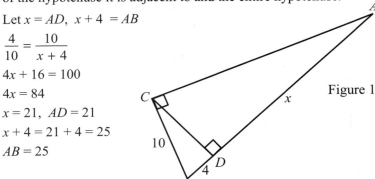

Let $x = AD$, $x + 4 = AB$

$\dfrac{4}{10} = \dfrac{10}{x + 4}$

$4x + 16 = 100$

$4x = 84$

$x = 21$, $AD = 21$

$x + 4 = 21 + 4 = 25$

$AB = 25$

Figure 1

❷ In a right triangle with the altitude drawn to the hypotenuse, the
length of the altitude is the mean proportional between the lengths
of the segments of the hypotenuse. $\dfrac{AD}{AC} = \dfrac{CD}{DB}$

If $CD = 6$ and BD is 9 less than AD. Find AB and DB.
Solution: The altitude is the mean proportional between the
segments it forms on the hypotenuse.

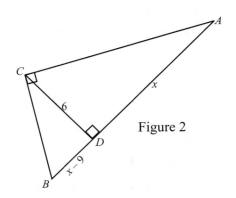

Let x = length of AD

$\dfrac{BD}{CD} = \dfrac{CD}{AD}$

$\dfrac{x - 9}{6} = \dfrac{6}{x}$

$x(x - 9) = 36$

$x^2 - 9x = 36$

$x^2 - 9x - 36 = 0$

$x = 12$ $x = -13$ reject

$x - 9 = 12 - 9 = 3$

$AD = 12$, $BD = 3$

Figure 2

Geometry Made Easy

Special Right Triangles:

There are two right triangles that are in many types of math problems. It is a good idea to memorize the basic two triangles and be able to sketch them on your scrap paper for reference when needed.

In a 45° - 45° - 90° right triangle: This is an isosceles right triangle. The hypotenuse is the product of either side and $\sqrt{2}$. The basic triangle is shown here. Multiples of the sides and hypotenuse are easily developed.

Examples

❶ Find the exact length, in inches, of one leg of an isosceles right triangle if the hypotenuse is $5\sqrt{2}$ inches.

Answer: Each leg is 5 inches.

❷ Find the length of the hypotenuse, to the nearest 10th of a cm, if one leg is 15 cm.

Answer: Hypotenuse = $15\sqrt{2} \approx 21.2$

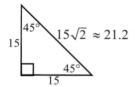

In a 30° - 60° - 90° right triangle: The side opposite the 30° angle is the length of the hypotenuse. The longer leg, opposite the 60° angle equals the product of the shorter leg and $\sqrt{3}$.

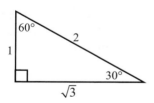

❶ The hypotenuse of 30°- 60°- 90° right triangle is 16cm. Find the exact lengths of both legs in centimeters.

Answer: The shorter leg = $\frac{1}{2}$ (16) = 8

The longer leg = $8\sqrt{3}$

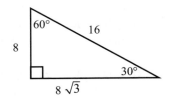

❷ The side opposite the 30 angle in a right triangle is 7.2 inches. Find the length of the hypotenuse and the longer leg, both to the nearest 10th of an inch.

Answer: Hypotenuse = 2(7.2) = 14.4

Longer leg = $7.2\sqrt{3}$ ≈ 12.5

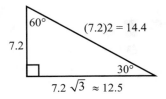

Special Right Triangles and the Pythagorean Theorem

The measures of the basic triangles can easily be found using the Pythagorean Theorem.

– 30°- 60°- 90°: Label the side opposite the 30° with one, make the hypotenuse two, and then use Pythagorean Theorem to find the longer leg.

– 45°- 45°- 90°: Start with a label of one for a leg opposite a 45°. Since it is isosceles, the other leg is also one. Then use Pythagorean Theorem to find the hypotenuse.

6 — QUADRILATERALS

Quadrilaterals - 4 sided polygons: The relationship between the various kinds of quadrilaterals is shown here. The characteristics of each quadrilateral are used in many types of geometry problems and proofs.

MEMORIZE THEM!

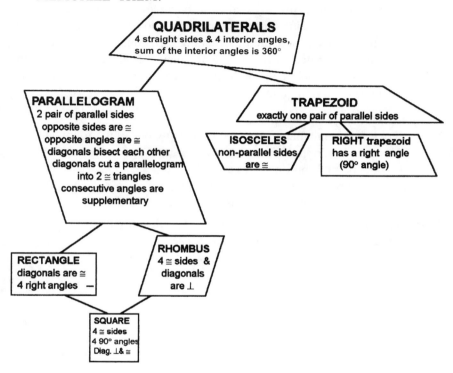

QUADRILATERALS
4 straight sides & 4 interior angles,
sum of the interior angles is 360°

PARALLELOGRAM
2 pair of parallel sides
opposite sides are ≅
opposite angles are ≅
diagonals bisect each other
diagonals cut a parallelogram
into 2 ≅ triangles
consecutive angles are
supplementary

TRAPEZOID
exactly one pair of parallel sides

ISOSCELES
non-parallel sides
are ≅

RIGHT trapezoid
has a right angle
(90° angle)

RECTANGLE
diagonals are ≅
4 right angles —

RHOMBUS
4 ≅ sides &
diagonals
are ⊥

SQUARE
4 ≅ sides
4 90° angles
Diag. ⊥ & ≅

PARALLELOGRAM

This section provides "reasons" that can be applied to various kinds of quadrilaterals. We will work with two subsets of quadrilaterals - parallelograms and trapezoids.

A parallelogram is a subset of the set of quadrilaterals. It has many characteristics that can be used to prove the figure is a parallelogram. If the quadrilateral is already known to be a parallelogram, these facts can be used to develop further geometric relationships in the subsets of parallelograms which include rectangles, rhombuses (sometimes the plural is written as rhombi), and squares. In order to work with any of the subsets of parallelograms, it is necessary to *prove that the figure is, first of all, a parallelogram.*

1) Known: A quadrilateral is given.
   Show: It is a **parallelogram** by proving any one of the following:
   – one pair of opposite sides are both parallel and congruent.
   – both pair of opposite sides are parallel.
   – both pair of opposite sides congruent.
   – the diagonals of the quadrilateral bisect each other.

Note: A quadrilateral must be proven to be a parallelogram before continuing with work involving the subsets.

Examples Prove a quadrilateral is a parallelogram

❶ Given: Parallelogram $BHDG$, $\overline{AH} \cong \overline{GC}$
   Prove: $ABCD$ is a parallelogram

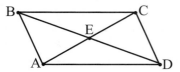

Statement	Reason
1. Parallelogram $BHDG$, $\overline{AH} \cong \overline{GC}$	1. Given
2. $\overline{BH} \cong \overline{DG}$, $\overline{BH} \parallel \overline{DG}$ (making $\overline{AB} \parallel \overline{DC}$)	2. Opposite sides of a parallelogram are congruent and parallel.
3. $\overline{BH} + \overline{HA} = \overline{DG} + \overline{GC}$	3. Addition Axiom
4. $\overline{BH} + \overline{HA} = \overline{AB}$ $\overline{DG} + \overline{GC} = \overline{CD}$	4. Segment Addition
5. $\overline{AB} \cong \overline{CD}$	5. Substitution
6. $ABCD$ is a parallelogram	6. If a quadrilateral has one pair of opposite sides that are both parallel and congruent, it is a parallelogram.

❷ Given: $\triangle AEB \cong \triangle CED$
   Prove: $\triangle ABCD$ is a parallelogram

Statement	Reason
1. $\triangle AEB \cong \triangle CED$	1. Given
2. $\angle DBA \cong \angle BDC$, $\overline{AB} \cong \overline{CD}$	2. CPCTC
3. $\overline{AB} \parallel \overline{CD}$	3. If two lines are cut by a transversal and alternate interior angles are ≅, the lines are parallel.
4. $ABCD$ is a parallelogram	4. If one pair of sides in a quadrilateral is both parallel and congruent, the figure is a parallelogram.

<u>Note:</u> There are many other ways to do this proof.

One other method would be to use CPCTC to show that the diagonals bisect each other, if so then, that would be proof of a parallelogram, also.

2) Known: A quadrilateral is a parallelogram.
Show: It is a **rectangle**. (All rectangles are parallelograms, so the proof must use something specific to the rectangles.)
   - A rectangle is a parallelogram that contains one right angle.
     [If it has one right $\angle$, then all of the $\angle$'s are right]
   - A parallelogram with diagonals that are $\cong$ is a rectangle.

Example

Given: Parallelogram $ABCD$, diagonals
$\overline{AC}$ and $\overline{BD}$ intersect at E.

$\triangle BEA \cong \triangle DEC, \angle BAE \cong \angle ABE$

Prove: $ABCD$ is a rectangle

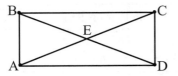

Statement	Reason
1. Parallelogram $ABCD$, diagonals $\overline{AC}$ and $\overline{BD}$ intersect at E. $\triangle BEA \cong \triangle DEC, \angle BAE \cong \angle ABE$	1. Given
2. $\overline{AE} \cong \overline{EC}$ and $\overline{BE} \cong \overline{ED}$	2. *CPCTC*
3. $\overline{AE} \cong \overline{BE}$	3. In a $\triangle$, if 2 angles are $\cong$ the sides opposite those angles are $\cong$.
4. $\overline{AE} \cong \overline{ED}$ and $\overline{BE} \cong \overline{EC}$	4. Transitive Property of equality
5. $\overline{AE} + \overline{EC}$ and $\overline{BE} + \overline{ED}$	5. Addition Property of equality
6. $\overline{AE} + \overline{EC} \cong \overline{AC}$	6. The whole is equal to the sum of its parts. (Or segment addition or the Partition Postulate.)
7. $\overline{AC} \cong \overline{BD}$	7. Substitution
8. $ABCD$ is a rectangle	8. A parallelagram with congruent diagonals is a rectangle.

Another method - Both pair of opposite sides can be proven to be parallel by proving the $\triangle AED \cong \triangle CEB$, then use the alternate interior angle information as we did in the sample proof for both pair of sides..

One more... Again if you prove that $\triangle AED \cong \triangle CEB$, you can use that with the congruent parts as shown in the sample proof to prove both pair of opposite sides are congruent.

Show: A parallelgram is a square – See page 58

Geometry Made Easy

RHOMBUS

Show: It is a **rhombus**. (Again, it has everything a parallelogram has, and specific proof is needed).

- A parallelogram with two adjacent sides that are ≅ is a rhombus.
- A parallelogram with perpendicular diagonals is a rhombus.
- A parallelogram with diagonals that bisect the angles of the parallelogram is a rhombus.

Examples Rhombus Examples

❶ In rhombus $ABCD$ the measure in centimeters of $\overline{AB}$ is $3x + 2$ and $\overline{BC}$ is $2x + 9$.
Find the number of centimeters in the length of $\overline{DC}$.

Solution: A rhombus has 4 ≅ sides.

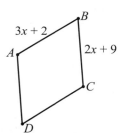

Make an equation: $2x + 9 = 3x + 2$

$7 = x$

$3(7) + 2 = 23$

$AB = 23$

$\overline{AB} \cong \overline{DC}$

$CD = 23$

❷ The diagonals of a rhombus are 10 and 16 inches in length.
Find the length of each side.

Solution: The diagonals are ⊥ and bisect each other. This forms right triangles and we can use the Pythagorean Theorem to solve the problem. The sides of the rhombus are all congruent.

$AE = 10/2 = 5$ inches

$BE = 16/2 = 8$ inches

$\overline{AB}$ is the hypotenuse in $\triangle AEB$ (c)

$c^2 = 5^2 + 8^2$

$c^2 = 89$

$c = \sqrt{89}$

Therefore, each side of the rhombus measures $\sqrt{89}$ inches.

Note: Since the problem did not say how to round, leave the answer in simplest radical form.

Rhombus Statement Reason Proof

Given: $\angle BAC \cong \angle BCA$, $\angle DAC \cong \angle DCA$,
$\angle ABD \cong \angle ADB$, $\angle CBD \cong \angle CDB$.

Prove: $ABCD$ is a rhombus

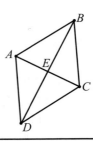

Statement	Reason
1. $\angle BAC \cong \angle BCA$, $\angle DAC \cong \angle DCA$, $\angle ABD \cong \angle ADB$, $\angle CBD \cong \angle CDB$.	1. Given
2. $\angle BAC + \angle DAC \cong \angle BCA + \angle DCA$ $\angle ABD + \angle CBD \cong \angle ADB + \angle CDB$	2. Addition Axiom. (When equals are added to equals, the results are equal.)
3. $\angle BAC + \angle DAC = \angle BAD$ $\angle BCA + \angle DCA = \angle BCD$ $\angle ABD + \angle CBD = \angle ABC$ $\angle ADB + \angle CDB = \angle ADC$	3. Angle addition postulate. (The whole is equal to the sum of its parts.)
4. $\angle BAD \cong \angle BCD$ $\angle ABC \cong \angle ADC$	4. Substitution
5. $ABCD$ is a parallelogram	5. In a quadrilateral, if both pairs of opposite angles are congruent, it is a parallelogram.
6. In $\triangle ABD$, $\overline{AD} \cong \overline{AB}$	6. In a triangle, if 2 angles are $\cong$, the sides opposite them are $\cong$.
7. $ABCD$ is a rhombus	7. A parallelogram with 2 consecutive (or adjacent) sides that are congruent is a rhombus.

Show: It is a **square** (combines all the characteristics of rectangles and rhombuses).

- A parallelogram with a right angle and whose two consecutive sides are $\cong$ is a square..
- A parallelogram whose diagonals are $\cong$ and $\perp$ to each other is a square.
- A rhombus with $\cong$ diagonals is a square.

Geometry Made Easy

TRAPEZOIDS

Trapezoids: a subset of quadrilaterals, have *exactly one pair* of parallel sides. Subsets of trapezoids include: <u>Isosceles trapezoid</u>, <u>right trapezoid</u>. It is necessary to prove the quadrilateral is a trapezoid before continuing to work on it as an isosceles or a right trapezoid.

Known: A quadrilateral is given.

Show: It is a **trapezoid** by proving
- One pair of sides is parallel and the other pair is not parallel.

Show: A trapezoid is an **isosceles trapezoid.**
- If the opposite non-parallel sides in a trapezoid are congruent, the trapezoid is isosceles.
- A trapezoid with congruent base angles is isosceles.

Additional information about Trapezoids:
- The **median** of a trapezoid
    a) Is parallel to the bases to the bases of the trapezoid.
    b) Has a length equal to half the sum of the length of the two bases.

Examples TRAPEZOID Examples

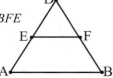

❶ Given: $\triangle ABD$, $\overline{EF}$ drawn to form trapezoid $ABFE$
$\angle DAB \cong \angle DEF$
$\overline{DE} \cong \overline{DF}$
Prove: $ABFE$ is an isosceles trapezoid.

Statement	Reason
1. $\triangle ABD$, drawn to form trapezoid $ABFE$ $\angle DAB \cong \angle DEF$ $\overline{DE} \cong \overline{DF}$	1. Given
2. $\overline{AB} \parallel \overline{EF}$	2. When 2 lines are cut by a transversal and corresponding angles are $\cong$, the lines are $\parallel$
3. $\angle DEF \cong \angle DFE$	3. In a triangle if 2 sides are $\cong$, the angles opposite them are $\cong$
4. $\angle DAB \cong \angle DFE$	4. Transitive Property
5. $\angle DFE \cong \angle DBA$	5. Corresponding $\angle$'s are $\cong$
6. $\angle DAB \cong \angle DBA$	6. Transitive Property
7. $ABFE$ is an isosceles trapezoid	7. If a pair of base angles in a trapezoid are congruent, the trapezoid is isosceles.

❷ In isosceles trapezoid *KJLM*, find the measures of each of the 4 angles.

A) $m\angle KML = 2(2x + 10)$
and $m\angle JLM = 5(x - 1)$

Solution: Since the trapezoid is isosceles, base angles are congruent.

$$2(2x + 10) = 5(x - 1)$$
$$4x + 20 = 5x - 5$$
$$25 = x$$
$$m\angle KML = m\angle JLM = 5(25 - 1) = 120$$
($\angle KML$ and $\angle MKJ$ are supplementary angles.)
$$m\angle MKJ = m\angle KJL = 180 - 120 = 60$$

❸ Given: Isosceles Trapezoid *KJLM*

If $ML = 12$, $KM = 10$, $KJ = 22$, find the length of the altitude, $\overline{ME}$, to $\overline{KJ}$ drawn from *M*.

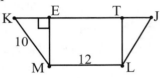

Solution: Drawn another altitude, $\overline{LT}$ from *L* to $\overline{KJ}$. Since $\overline{ML} \parallel \overline{KJ}$ and $\overline{LT}$ is perpendicular to them both, $\overline{ML} \cong \overline{ET}$. *KJLM* is isosceles, so it is symmetrical. Since *KJ*= 22 and $\overline{ML}$ = 10, the 2 segments left over are each 5: *KE* = 5, *TJ* = 5. $\overline{KM}$ is 10 and it is the hypotenuse of right ΔKEM. Use the Pythagorean Theorem to find the height,

$$5^2 + (ME)^2 = 10^2$$
$$(ME)^2 = 75 \qquad \text{The length of the altitude is } 5\sqrt{3}.$$
$$ME = \sqrt{75} = 5\sqrt{3}$$

<u>Note:</u> Since the length of one leg of this right triangle is ½ the length of the hypotenuse, it is clearly a "special right triangle" with angles of 30°, 60°, 90°. That makes the shorter leg of the triangle automatically equal to half the length of the hypotenuse multiplied by $\sqrt{3}$. (See page 51 – Special Right Triangles)

7 — CONCURRENCE IN TRIANGLES

Medians, Altitudes, ⊥ bisectors, angle bisectors and related information

Concurrence: If three or more lines share a single point of intersection, the lines are concurrent and the point of intersection is the point of concurrency.

In a triangle the medians are concurrent, the altitudes are concurrent, the angle bisectors are concurrent, and the perpendicular bisectors of the sides are concurrent. Each of these four points of intersection has a special name. Sometimes the point of concurrency is outside the triangle.

Quick review of terms:

- Median – connects a vertex with the midpoint of the opposite side of the triangle. $\overline{CD}$ is the median drawn to $\overline{AB}$.

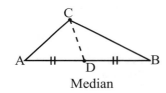

Median

- Altitude – a line perpendicular to a side of the triangle and drawn to the opposite vertex. $\overline{CE}$ is the altitude to base $\overline{AB}$.

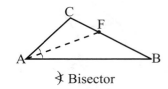

Altitude

- Angle Bisector – A line drawn from a vertex to the opposite side which divides the angle into two congruent angles. $\overline{AF}$ bisects ∢CAB.

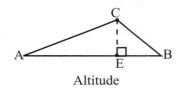

∢ Bisector

- Perpendicular bisector – A line perpendicular to a side of the triangle at its midpoint. $\overline{DG}$ is the ⊥ bisector of side $\overline{AB}$.

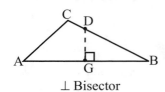

⊥ Bisector

Circles Related To Triangles:

Circumscribed Circle: A circle circumscribed about a polygon goes through all the vertices of the polygon. The polygon is completely inside the circle. See page 64.

Inscribed Circle: A circle inscribed in a polygon is tangent to all the sides of the polygon. (In a regular polygon, the radius of an inscribed circle is called the apothem.) The circle is completely inside the polygon. See diagram below

Concurrence – Diagrams and Theorems:

• **Incenter:** The point of concurrency of the three angle bisectors of a triangle. The incenter is inside the triangle.

Theorem: The incenter of a triangle is equidistant from each side of the triangle.

Incenter Diagram

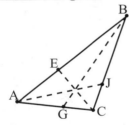

$\overline{AJ}$ bisects $\angle BAC$, $\angle BAJ \cong \angle CAJ$

$\overline{EC}$ bisects $\angle ACB$, $\angle ACE \cong \angle BCE$

$\overline{BG}$ bisects $\angle ACB$, $\angle ABG \cong \angle CBG$

Inscribed Circle: A circle can be inscribed in a triangle by using the incenter as its center and the perpendicular distance to a side as the radius.

Inscribed Circle

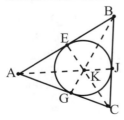

K is the incenter of $\triangle ABC$. Circle K is inscribed in $\triangle ABC$. The radius is the perpendicular distance from K to any side of $\triangle ABC$.

- **Centroid:** The point of concurrency (intersection) of the 3 ***medians*** of a triangle.

  Theorem: The centroid of a triangle is two thirds of the distance from a vertex to the side opposite that vertex. (The distance from the centroid to the vertex is two times the distance from the centroid to the side. The ratio is 2:1.)

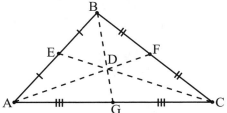

$$FD : DA = 1 : 2 \text{ or } \frac{FD}{DA} = \frac{1}{2}$$

$$CD = \frac{2}{3} CE$$

If $BD = 10$, then $DG = 5$

Note: The centroid is the center of balance for the triangle – if the triangle were cut out of stiff cardboard, it could be balanced on a pencil point at the centroid.

- **Orthocenter:** The point of concurrency of the three altitudes of a triangle. The orthocenter can be inside or outside a triangle.

Figure 1

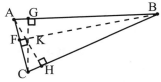

$$\overline{AH} \perp \overline{BC}$$
$$\overline{CG} \perp \overline{AB}$$
$$\overline{BF} \perp \overline{AC}$$

K is the internal orthocenter

Figure 2

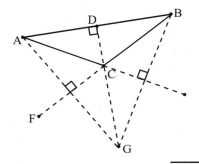

Sides $\overline{BC}$ and $\overline{AC}$ are extended.
$\overline{DY}$ is the altitude to side $\overline{AB}$,
$\overline{BG}$ is the altitude to side $\overline{AC}$
$\overline{AG}$ is the altitude to side $\overline{BC}$
G is the external orthocenter.

- **Circumcenter:** The point of concurrency created by the three perpendicular bisectors of the three sides of a triangle. The circumcenter can be inside (Figure 1) outside the triangle (Figure 2).

Theorem: The circumcenter of a triangle is equidistant from the three vertices of the triangle.

(Figure 1)

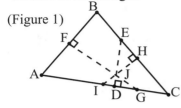

$$\overline{ED} \perp \overline{AC} \ , \ \overline{AD} \cong \overline{DC}$$

$$\overline{FG} \perp \overline{AB} \ , \ \overline{AF} \cong \overline{FB}$$

$$\overline{IH} \perp \overline{BC} \ , \ \overline{BH} \cong \overline{HC}$$

J is the circumcenter.

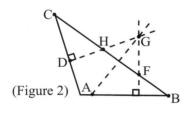

(Figure 2)

$$\overline{DG} \perp \overline{CA} \ , \ \overline{CD} \cong \overline{AD}$$

$$\overline{AG} \perp \overline{BC} \ , \ \overline{CH} \cong \overline{BH}$$

$$\overline{FG} \perp \overline{AB} \ , \ \overline{AF} \cong \overline{BF}$$

G is the circumcenter.

Circumscribed Circle

A circle can be circumscribed about a triangle by using the circumcenter as its center and the distance from the circumcenter to a vertex as the radius.

Circumscribed Circle Diagrams

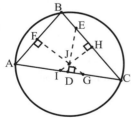

J is the circumcenter of △*ABC*.
Circle *J* is circumscribed about △*ABC*.
The radius is the distance from *J* to any any vertex of the triangle, *A*, *B*, or *C*.

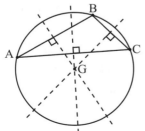

G is the circumcenter of △*ABC*. Circle *G* is circumscribed about △*ABC*. The radius of *G* is the distance from *G* to any vertex of the triangle.

Note: A circle can be circumscribed about a triangle by using the circumcenter as its center and the distance from the circumcenter to a vertex as the radius.

Geometry Made Easy

8 — ANALYTIC GEOMETRY OR COORDINATE GEOMETRY

Each point on a plane corresponds to one specific location named by the appropriate ordered pair of numbers in the form (x, y). A rectangular coordinate system is usually shown on a "grid" that represents the plane. The plane is also called a Cartesian plane. The "x" value of an ordered pair is called the **abscissa**. The "y" value is called the **ordinate**. The correspondence of each point in a plane to an ordered pair in a coordinate system allows us to locate points on the plane. Using a "grid" or coordinate graph we can use algebraic laws and processes along with the geometry "reasons" shown previously to solve problems. The horizontal axis of the graph is the x-axis and the vertical axis is the y-axis. The coordinate graph is understood to have a scale of 1:1 unless labeled otherwise. The intersection of the two axes at $(0, 0)$ is called the **origin**. The x-axis and y-axis are accepted as being perpendicular to each other. (Remember to label the axes, the origin, the scale, and to put arrows on the ends of the lines representing the axes on all graphs.)

_Y_-Intercept: This is the "y" value at the point where the graphed line crosses the y-axis. It can be found by substituting "0" for "x" in the given equation and finding the value for "y". (Remember that everywhere on the y-axis, $x = 0$.) The y-intercept point is often used as a starting point when graphing equations from the slope intercept form of an equation.)

Example Find the y-intercept of the equation $2y = 4x + 12$.

Substitute: $2y = 4(0) + 12$
$2y = 12$
$y = 6$
The y-intercept is 6.
This means the graph crosses the y-axis at $(0, 6)$

_or_ Use the slope-intercept form of the equation, $y = mx + b$.

Isolate "y", and set the equation up in the form $y = mx + b$.

"b" is the y-intercept. Since $b = 6$, the point the line crosses the y-axis is $(0,6)$.

Slope: Slope tells the "steepness and direction" of the slant of a line on a graph. It is the ratio of the vertical change in the line to the horizontal change in the line. (In science classes, slope is often described as "rise over run.") Slope is written as a fraction. Although "*m*" is used to represent the slope in formulas, it is necessary to write the word "slope" when the final answer to the formula is found or when slope is used in other ways. Use the slope to find points on the line. The top (numerator) of the fraction tells the vertical movement along the line from one point to another. The bottom (denominator) of the fraction shows the horizontal movement along the line between the points.

Example   Negative slope

$$-\frac{3\downarrow}{4\rightarrow}$$

4 units to the right,
3 units down.

Positive Slope

$$\frac{3\uparrow}{4\rightarrow}$$

4 units to the right,
3 units up.

Hint: The "upper" part of the slope fraction tells the "up or down" movement on the graph. Write arrows on your slope fraction to help you remember.

1) **Slope And Equations:**

Slope formula: $m = \dfrac{y_2 - y_1}{x_2 - x_1}$ where (x_2, y_2) is one point on the line and (x_1, y_1) is another point on the line.

Example   Find the slope of the line through (4, 7) and (−2, −3)

$$m = \frac{-3 - 7}{-2 - 4} = \frac{-10}{-6} = \frac{5}{3} \text{ (over 3 and up 5 when used for graphing).}$$

2) **Slope - Intercept:** Slope can also be found as the coefficient of "*x*" when a linear equation is written in the form $y = mx + b$. "*m*" represents the slope.

Example $y = 5x - 4$. Slope is 5 or $\dfrac{5}{1}$.
(1 unit right and 5 units up when graphing)

Geometry Made Easy

3) Point-Slope Form of an equation: To write the equation of a line, if you know the slope and a point on the line, use this formula: $y - b = m(x - a)$ where "m" is the slope of the line and (a, b) is a point on the line.

Example Write the equation of a line going through the point $(-4, 5)$ and having a slope of $\frac{3}{4}$.

Steps

1) Substitute. $\qquad\qquad y - 5 = \frac{3}{4}(x - (-4))$

2) Multiply. $\qquad\qquad y - 5 = \frac{3}{4}(x + 4)$

$$y - 5 = \frac{3}{4}x + 3$$

3) Isolate "y". $\qquad\qquad \underline{+\ 5 \qquad\quad +\ 5}$

4) Answer. $\qquad\qquad y = \frac{3}{4}x + 8$

Special Slopes: MEMORIZING THEM is best.

Vertical Lines: Have no slope (slope is undefined).

Example $x = 3$ is a vertical line crossing the x-axis at 3. It is parallel to the y-axis. 2 points on the line are $(3, 0)$ and $(3, 2)$. Using $m = \frac{\Delta y}{\Delta x}$ causes the denominator to be zero. The slope is undefined since division by zero is undefined.

Horizontal Lines: Have a slope $= 0$.

Example $y = -4$ is a horizontal line crossing the y-axis at -4. It is parallel to the x-axis. $y = -4$ goes through $(0, -4)$ and $(5, -4)$. $m = \frac{\Delta y}{\Delta x}$; so the numerator of the slope fraction is zero. Therefore, the slope $= 0$.

Collinear points: Two or more points that are on the same line. If the slope is the same between the first and second points as it is between the second and third (or first and third), then the points are on the same line. Use the slope formula to test this.

(Example of Collinear points on next page)

Example Are (3, 5), (6, 8), and (−4, −8) collinear points?

Solution: Call (3, 5) point A, (6, 8) point B, and (−4, −8) point C. First, find the slope between points A and B. Then use the same formula for points B and C.

Slope of points 1 and 2

$$m = \frac{8-5}{6-3} = \frac{3}{3} = 1$$

Slope between points 2 and 3

$$m = \frac{-8-8}{-4-6} = \frac{-16}{-10} = \frac{8}{5}$$

Since the slope between the first two points is 1, and the slope between the second and third points is $\frac{8}{5}$, thus not the same, these three points are *not* on the same line.

Parallel and Perpendicular Lines:

Mathematically correct thinking processes must be used in coordinate geometry to analyze the relationship between lines. These are some that are used with the slope formula or the equation form of the line ($y = mx + b$) to prove whether or not lines are parallel ($\parallel$) or perpendicular ($\perp$).

- If two lines have the <u>same slope</u>, the lines are parallel.

  Example Are $y = 3x − 4$ and $y = 3x + 7$ parallel lines? They both have a slope of 3, so yes, they are parallel lines.

- If two lines are <u>parallel</u>, they have the same slope.

  Example If a line has a slope of −3, write the equation of another line that is parallel to it: $y = −3x + 1$. Both lines have a slope = −3, so they are parallel.

- If two lines have slopes that are <u>negative reciprocals</u> of each other, the lines are perpendicular.

  Example Are these two lines perpendicular?
  $y = 3x − 4$ and $y = −\frac{1}{3}x + 7$. Since the slopes are 3 and $−\frac{1}{3}$ which are negative reciprocals of each other, the lines are perpendicular.

- If two lines are <u>perpendicular</u>, their slopes are negative reciprocals of each other.

  Example Name two points that would be on a line that is perpendicular to the line $y = −2x$. To solve this, we must find two points that will work in the slope formula to make a slope of $\frac{1}{2}$. An answer might be (6, 3) and (12, 6).

Note: When two numbers that are negative reciprocals of each other are multiplied together, their product is −1.

Geometry Made Easy

Writing the Equation for parallel or perpendicular lines

- Write the equation of a line perpendicular to a given line through a given point on the line: Find the slope of the original line. The slope of the line perpendicular to it is the negative reciprocal of that number. The x and y values of the point are used to find the y-intercept of the new line.

Example Find the equation of a line perpendicular to the line $y = 2x + 3$ that goes through the point $(6, 9)$.

Solution: The slope of the given line is 2. Therefore the slope of a line perpendicular to it will be $(-1/2)$. Use $(-1/2)$ in the slope intercept form of an equation and substitute the values $(6, 9)$ in it to find the value of b. Rewrite the new equation using x and y.

$$9 = -\frac{1}{2}(6) + b$$
$$9 = -3 + b$$
$$b = 12$$

Answer: $y = -\frac{1}{2}x + 12$

- Write the equation of a line that is the perpendicular bisector of a line segment. Same procedure as the above example but we must find the midpoint of the segment first.

Example Find the equation of the line that is the perpendicular bisector of the line segment joining the points $(-2, -4)$ and $(4, -6)$.

Solution: Find the slope of the given line segment. Find the midpoint of the given segment. Use the same method we used in the above example to write the equation.

Slope of given line

$$m = \frac{y_2 - y_1}{x_2 - x_1}$$

$$m = \frac{-6 - (-4)}{-4 - (-2)}$$

$$m = -\frac{1}{3}$$

Midpoint of given segment

$$M = \left(\frac{x_1 + x_2}{2}, \frac{y_1 + y_2}{2} \right)$$

$$M = \left(\frac{-2 + 4}{2}, \frac{-4 + (-6)}{2} \right)$$

$$M = (1, -5)$$

The slope of the perpendicular bisector will be 3. It will go through the midpoint, $(1, -5)$

$$-5 = 3(1) + b$$
$$b = -8$$

Answer: $y = 3x - 8$

- Find the equation of a line parallel to another line through a specific point. This is similar to the first example on page 69, although the new equation must have the same slope as the given line.

Example Write the equation of a line parallel to line $y = 3x - 5$ that goes through the point $(-3, 4)$

Slope of the given line is 3. The slope of the new line is also 3. Substitute $(-3, 4)$ to find b.

$$4 = 3(-3) + b$$
$$b = 13$$

Answer: $y = 3x + 13$ ans.

When developing equations of lines parallel or perpendicular to another line, be careful to pay attention to the slope needed. Sketch the problem on graph paper or check it in your graphing calculator to see that the equations do actually produce lines that are parallel or perpendicular to each other.

Length of a segment: Finding the length of a segment is necessary to prove whether segments are congruent or not. The length is also needed to find the area of geometric figures on a graph. If the segment is not horizontal or vertical, the distance formula is needed.

A horizontal segment is *parallel to the x-axis* and has a length equal to the absolute value of the difference of the "*x*" values of the end points of the segments. Length $= |x_2 - x_1|$.

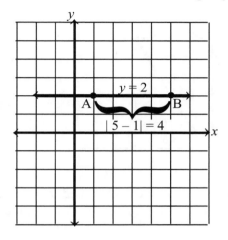

$A(1, 2)$
$B(5, 2)$
$AB = |5 - 1| = 4$

A vertical segment is *parallel to the y-axis* and has a length equal to the absolute value of the difference of the "*y*" values of the endpoints of the segments. Length $= |y_2 - y_1|$.

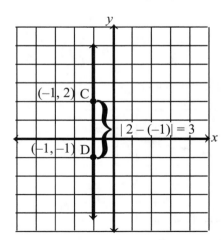

$C(-1, 2)$
$D(-1, -1)$
$CD = |2 - (-1)| = 3$

Distance Formula: This formula is used to find the length of a line segment that is *not parallel* to either axis. Using the (x, y) values for both endpoints, the formula is $d\sqrt{(x_2 - x_1)^2 + (y_2 - y_1)^2}$.

Example Find the length of the line segment joining the points $(4, 6)$ and $(1, 2)$. Consider $(4, 6)$ to be point 1. $(1, 2)$ is then point 2.

$$d\sqrt{(1 - 4)^2 + (2 - 6)^2} = \sqrt{(-3)^2 + (-4)^2} = \sqrt{9 + 16} = \sqrt{25} = 5$$

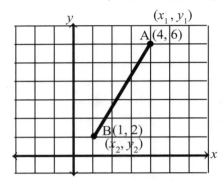

Midpoint of a line segment: The midpoint is used to show lines are congruent, to show that a line is bisected by another line, and to assist in proving many kinds of problems. The coordinates of the midpoint of a line segment are found by using the following formula with the x and y values of the endpoints of the segment.

Midpoint Formula: $(x, y)_{midpoint} = \left(\dfrac{x_2 + x_1}{2}, \dfrac{y_2 + y_1}{2} \right)$

Example Find the coordinates of the midpoint of the segment joining $(-2, 4)$ and $(8, 6)$.

$$x = \frac{-2 + 8}{2} \Rightarrow \frac{6}{2} \Rightarrow 3 \quad \text{and} \quad y = \frac{4 + 6}{2} \Rightarrow \frac{10}{2} \Rightarrow 5$$

The coordinates of the midpoint are $(3, 5)$.

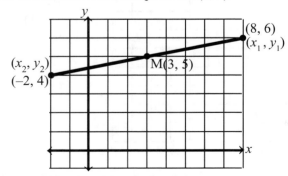

Geometry Made Easy

Coordinate or Analytic Proof Examples: Use the slope, distance, and midpoint formulas to prove the relationships between the sides of a given figure to make a conclusion. Several examples follow, but remember that there are usually several ways to solve each one. *Your work must be done with your textbook and your own teacher's instructions in mind.*

   Note: Good organization and clearly written conclusions are necessary .

Examples

① Given the quadrilateral $ABCD$ with its vertices at $A(3, -5)$, $B(5, 1)$, $C(1, 4)$ and $D(-5, 1)$. Determine (prove) whether or not $ABCD$ is an isosceles trapezoid.

To solve, draw a diagram on graph paper (see to the right) and make a plan for your work: First prove that the diagram is a trapezoid by showing that exactly two sides are parallel using the slope formula. $\overline{CB}$ is parallel to $\overline{DA}$ and $\overline{CD}$ is not parallel to $\overline{AB}$. Then to isosceles, the nonparallel sides must be equal. Use the distance formula to find this out. Find the length of $\overline{CD}$ and then the length of $\overline{AB}$ and compare.

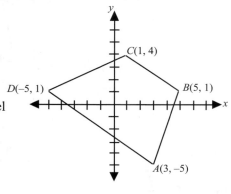

Slope formula: $m = \dfrac{y_2 - y_1}{x_2 - x_1}$

Distance Formula: $d = \sqrt{(x_2 - x_1)^2 + (y_2 - y_1)^2}$

$m_{\overline{CB}} = \dfrac{4 - 1}{1 - 5} = \dfrac{3}{-4} = -\dfrac{3}{4}$

$d_{\overline{CD}} = \sqrt{(1 - (-5)^2 + (4 - 1)^2}$

$m_{\overline{DA}} = \dfrac{1 - (-5)}{-5 - 3} = \dfrac{6}{-8} = -\dfrac{3}{4}$

$d_{\overline{CD}} = \sqrt{(1 + 5)^2 + (3)^2} = \sqrt{36 + 9} = \sqrt{45}$

$m_{\overline{CD}} = \dfrac{4 - 1}{1 - (-5)} = \dfrac{3}{6} = \dfrac{1}{2}$

$d_{\overline{AB}} = \sqrt{(5 - 3)^2 + (1 - (-5)^2}$

$m_{\overline{AB}} = \dfrac{-5 - 1}{3 - 5} = \dfrac{-6}{-2} = 3$

$d_{\overline{AB}} = \sqrt{(2)^2 + (1 + 5)^2} = \sqrt{4 + 36} = \sqrt{40}$

CONCLUSION: $\overline{CB} \parallel \overline{DA}$ because they have equal slopes. $\overline{CD}$ is not parallel to $\overline{AB}$ because their slopes are not equal. Therefore, quadrilateral $ABCD$ is a trapezoid because it has exactly one pair of parallel sides. $\overline{CD}$ has a length of $\sqrt{45}$ and $\overline{AB}$ has a length of $\sqrt{40}$ showing that the two nonparallel sides are not equal. Therefore, trapezoid $ABCD$ is *not* isosceles.

❷ Given: Quadrilateral $ABCD$ with vertices at $A(-4, 2)$, $B(2, 6)$, $C(6, 0)$ and $D(0, -4)$ and diagonals $\overline{AC}$ and $\overline{BD}$ which intersect at E.

Prove: $ABCD$ is a rhombus.

Plan: 1) Prove opposite sides are parallel to show it is a parallelogram. Use slope formula.

2) Prove diagonals are perpendicular - slopes are negative reciprocals of each other; or prove two adjacent sides are equal in length using the distance formula.

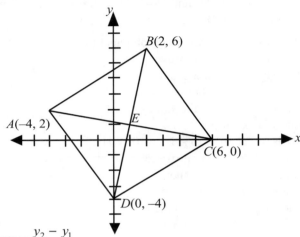

Slope Formula: $m = \dfrac{y_2 - y_1}{x_2 - x_1}$

Slope $\overline{AB} = \dfrac{6 - 2}{2 - (-4)} \Rightarrow \dfrac{4}{6} \Rightarrow \dfrac{2}{3}$ Slope $\overline{DC} = \dfrac{-4 - 0}{0 - 6} \Rightarrow \dfrac{-4}{-6} \Rightarrow \dfrac{2}{3}$

Slope $\overline{AD} = \dfrac{-4 - 2}{0 - (-4)} \Rightarrow \dfrac{-6}{4} \Rightarrow -\dfrac{3}{2}$ Slope $\overline{BC} = \dfrac{6 - 0}{2 - 6} \Rightarrow \dfrac{6}{-4} \Rightarrow -\dfrac{3}{2}$

Slope $\overline{AB}$ = Slope $\overline{DC}$ so $\overline{AB} \parallel \overline{DC}$, and slope $\overline{AD}$ = slope $\overline{BC}$, showing that $\overline{AD} \parallel \overline{BC}$. $ABCD$ is a parallelogram because opposite sides are parallel.

Slope $\overline{BD} = \dfrac{-4 - 6}{0 - 2} \Rightarrow \dfrac{-10}{-2} \Rightarrow 5$ Slope $\overline{AC} = \dfrac{2 - 0}{-4 - 6} \Rightarrow \dfrac{2}{-10} \Rightarrow -\dfrac{1}{5}$

Conclusion: 5 and $-\dfrac{1}{5}$ are negative reciprocals of each other. Therefore, $\overline{BD} \perp \overline{AC}$. $ABCD$ is a rhombus because it is a parallelogram with perpendicular diagonals.

---

9 — GRAPHIC SOLUTION OF QUADRATIC – LINEAR PAIRS OF EQUATION

Solving Quadratic Equations Graphically: Parabolas The graph of a quadratic equation in the form $y = ax^2 + bx + c$ where a, b, and c are real numbers and $a \neq 0$ is a parabola. The graph has a "**U**" shape. The position and exact shape of the "**U**" are determined by using the values of a, b and c in the equation. A parabola is the locus of points equidistant from a fixed line and a fixed point. The fixed point is called the *focus* and the fixed line is the *directrix*. In this book we will only consider parabolas which have a directrix that is parallel to the x-axis or the y-axis.

Vertical or Horizontal? Graphs from the equation $y = ax^2 + bx + c$ are vertical and graphs of $y = ay^2 + by + c$ are horizontal on a coordinate plane. We will work with vertical parabolas in the examples given. Horizontal parabolas have the same basic characteristics and the "vertical" information can easily be interpreted for them.

Up or Down? If "a" is negative, the graph of the parabola will open downward, if "a" is positive, the graph opens upward.

Y-Intercept: In the equation, $y = ax^2 + bx + c$, the constant, "c", is the "y" intercept.

Axis of Symmetry: The line of reflection of a parabola. Points on one side of the axis of symmetry are mirror images of the points on the other side. The axis of symmetry goes through the turning point of the parabola. To find the equation of the axis of symmetry use the formula $x = -b/2a$.

Turning Point or Vertex: The maximum (if parabola opens downward) or minimum (if the parabola opens upward) point on the graph. The turning point is on the axis of symmetry. The "x" coordinate of the turning point can be found by using $x = -b/2a$. The "y" coordinate of the turning point can then be found by substituting the value of "x" into the original equation, $y = ax^2 + bx + c$. The "x" coordinate of the turning point gives a "center" for the table of values used to graph a parabola.

Roots: The roots, or zeros, of the equation are the "x" values of the points where the parabola crosses the x-axis. There can be one, two, or no real roots. If it doesn't cross the x-axis at all, then there are no real roots. If it just touches the axis, there is one real root, and if it intersects it in two places, then there are two real roots. The roots can be found by reading the graph - accuracy in graphing is essential here. Approximate (estimate) the roots if they are not integers. [To check the roots: Since $y = 0$ at the roots, substitute 0 for "y" in the equation and substitute your answers for x.]

Graphing A Quadratic Equation

Steps

1) Solve the quadratic equation for "y" in terms of "x". It will be in the form $y = ax^2 + bx + c$. Make note of the values of a, b, and c.

2) Find the axis of symmetry: $x = \dfrac{-b}{2a}$. The vertex is on the axis of symmetry. Use the x value of the vertex as the middle value for the x's on the table.

3) Make a table of values for the parabola. Use at least three integral values on each side of the "x" coordinate of the vertex. (Try not to use fractions or decimals for the values of "x" you choose.) Sometimes the interval of the values for "x" to be used will be given.

4) Substitute the values of "x" to find the y-coordinate of each point. SHOW SUBSTITUTIONS.

5) Plot the points and sketch the graph accurately and with a smooth curve.

6) LABEL the curve. Label the roots and the turning point when possible.

Example

a) Solve by graphing: $y = x^2 + 4x - 5$ using the interval $(-5 \leq x \leq 1)$.

b) Write the equation for the axis of symmetry.

c) Write the roots as a solution set.

d) Locate its turning point and indicate if it is a maximum or minimum point.

(b) Axis of Symmetry

Root $x = -5$

Root $x = 1$

$x = -2$

$y = x^2 + 4x - 5$

(c) SS = {-5, 1}

(d) minimum turning point (-2, -9)

(a) x	$y = x^2 + 4x - 5$	y	(x, y)
-5	$y = (-5)^2 + 4(-5) - 5$	0	$(-5, 0)$
-4	$y = (-4)^2 + 4(-4) - 5$	-5	$(-4, -5)$
-3	$y = (-3)^2 + 4(-3) - 5$	-8	$(-3, -8)$
-2	$y = (-2)^2 + 4(-2) - 5$	-9	$(-2, -9)$
-1	$y = (-1)^2 + 4(-1) - 5$	-8	$(-1, -8)$
0	$y = (0)^2 + 4(0) - 5$	-5	$(0, -5)$
1	$y = (1)^2 + 4(1) - 5$	0	$(1, 0)$

(b) Axis of Symmetry

$$x = \frac{-b}{2a}$$

$$x = \frac{-(4)}{2(1)}$$

$$x = \frac{-4}{2}$$

$$x = -2$$

Geometry Made Easy

Quadratic - Linear Pair:

Two equations that are to be solved simultaneously, one is a second degree, the other is a first degree equation. In this level of math, the second degree equation would be a parabola or a circle on a graph and the first degree equation would be a straight line. There can be one, two or no solutions but frequently there are two.

Solving Graphically: Graph and label each equation on the same coordinate graph. The solution set of the system includes the point or points where the two graphs intersect each other. Use the method shown on page 76 to graph the quadratic equation. Then graph the linear equation using the form $y = mx + b$ where "m" is the slope and "b" is the y-intercept. Label their intersections and check the solution(s) in both original equations.

Example Solve the following system by graphing. Check.

$$y = x^2$$
$$y = 2x + 3$$

Show work:

$$y = x^2$$

$$x = \frac{-b}{2a}$$

$$x = \frac{0}{2}$$

$$x = 0$$

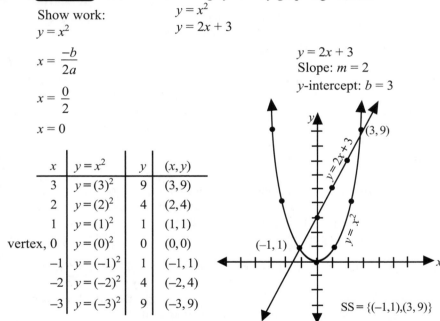

$y = 2x + 3$
Slope: $m = 2$
y-intercept: $b = 3$

x	$y = x^2$	y	(x, y)
3	$y = (3)^2$	9	$(3, 9)$
2	$y = (2)^2$	4	$(2, 4)$
1	$y = (1)^2$	1	$(1, 1)$
vertex, 0	$y = (0)^2$	0	$(0, 0)$
−1	$y = (-1)^2$	1	$(-1, 1)$
−2	$y = (-2)^2$	4	$(-2, 4)$
−3	$y = (-3)^2$	9	$(-3, 9)$

$SS = \{(-1, 1), (3, 9)\}$

Checking Systems of Equations: Always use the two original equations. The wrong answer will usually check in one but not in both equations.

Check:

$(-1, 1)$
$1 = (-1)^2$
$1 = 1 \checkmark$
$1 = 2(-1) + 3$
$1 = 1 \checkmark$

$(3, 9)$
$9 = 3^2$
$9 = 9 \checkmark$
$9 = 2(3) + 3$
$9 = 9 \checkmark$

10 — TRANSFORMATIONAL GEOMETRY

Transformational Geometry: A one-to-one mapping of points in a plane. Transformations are used to change the position of points (or a figure) in a plane. (See also pages 115–116)

Mapping: A mapping in transformation geometry pairs each point in the preimage with a point in its image based on the tranformation indicated. Symbol: $\longrightarrow$

Pre-Image: The original figure that is to be translated.

Image: The figure that is formed when each point in an original figure is moved in a specific way (mapped) which depends on the transformation being used.

Prime(′): A "prime", is used to associate the image point with its preimage point. So the image of point A is A'. For multiple transformations, double ($''$) and triple primes ($'''$) are used to show subsequent images.

Isometry: In a *direct isometry*, the distance between the points of a figure is maintained and the orientation of the figure maintained. The image is congruent to its pre-image. In an *indirect isometry,* or opposite isometry, distance is preserved but orientation is opposite when comparing the preimage and the image. Reflections, rotations, and translations are isometries.

Similarity: Dilations and composite transformations that contain a dilation are called similarities. The pre-image and its image are similar figures. (They have corresponding angles that are congruent and corresponding sides that are proportional.)

Invariant: Points or properties that do not change when the preimage is tranformed into its image. A point or points on the preimage and image may be unchanged or invariant. The distance, angle measures, betweeness and collinearity of points may be invariant as well.

Notation or Rule: The type of transformation required is indicated by using a letter to show the type of transformation and the additional information needed is given with it. There are examples shown in each description below. Sometimes the notation is done in front of the problem, sometimes it is over or under the arrow showing the mapping. The transformation itself can be written as a general rule using x and y and showing how they change as the transformation is performed.

Geometry Made Easy

Orientation (order): If the preimage has a clockwise direction of the labels on its points and the image has the same clockwise direction of its labels, they have the same or direct orientation. (Rotations, dilations, translations have the same orientation.) Opposite *or* indirect *or* reverse orientation *or* reverse order means that the pre-image and its image have opposite directions of their labels. If the pre-image is clockwise, the image is counter-clockwise. (Line reflections and glide reflections have opposite or reverse orientation.)

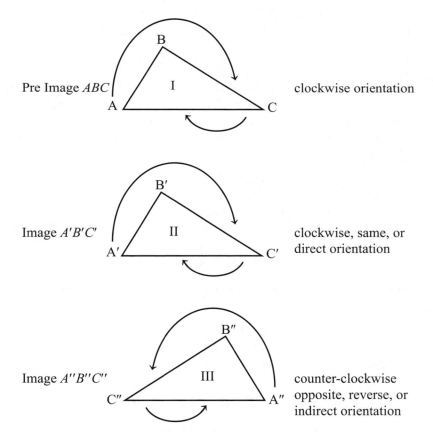

Pre Image *ABC*   clockwise orientation

Image *A'B'C'*   clockwise, same, or direct orientation

Image *A"B"C"*   counter-clockwise opposite, reverse, or indirect orientation

The preimage I and image II have the same orientation or direct orientation. Both are clockwise.

The preimage I and image III have reverse, opposite, or indirect orientation. One is clockwise, the other counter-clockwise.

Line Reflection (r): Under a line reflection on line "*m*", if a point, *A* is connected to its image, *A'*, a line segment *AA'* is formed and line "*m*" is the perpendicular bisector of *A*. In a line reflection, if the diagram is folded exactly on the line of reflection and the two halves are put together and held up to the light, the original figure and its image will be an exact match. Each side of the line of reflection is a "mirror image" of the other side. $\triangle A'B'C'$ is the image of $\triangle ABC$ under a line reflection line "*m*".

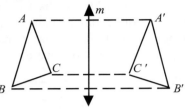

In symbols: $r_m(\triangle ABC) = \triangle A'B'C'$ _or_ $\triangle ABC \xrightarrow{r_m} A'B'C'$

Properties under a line reflection: Distance, angle measure, parallelism, collinearity are invariant. Any of these properties that are true for the preimage will remain unchanged in its image. Orientation in a line reflection is reversed or opposite. A line reflection is an opposite isometry.

Examples

❶ $r_{y\text{-axis}}$: If the line of reflection is the *y*-axis and the coordinates of the original point are (x, y), then the coordinates of its image will be $(-x, y)$.

$$(x, y) \xrightarrow{r_{y\text{-axis}}} (-x, y)$$

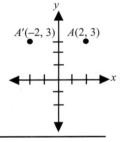

❷ $r_{x\text{-axis}}$: If the coordinates of the original point are (x, y), then the coordinates of its image will be $(x, -y)$.

$$(x, y) \xrightarrow{r_{x\text{-axis}}} (x, -y)$$

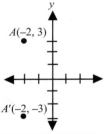

❸ $r_{y=x}$: If the line of reflection is the line representing the equation $y = x$ and the original point is (x, y), then its image is (y, x).

$$(x, y) \xrightarrow{r_{y=x}} (y, x)$$

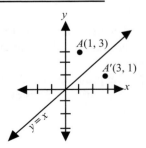

Geometry Made Easy

The line of reflection can be outside a figure or it may be part of the figure. It can be a line of symmetry when it is inside the figure. This one is part of the figure:

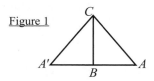

Figure 1

In this case, B and C are invariant points. They are their own images because they are on the line of reflection.

$$\triangle ABC \xrightarrow{r_{\overline{BC}}} \triangle A'B'C$$

Note: **R** is sometimes used instead of *r* in reflections. Since the **R** for a rotation will include an angle reference it is possible to distinguish between the two **R**'s. Follow your teacer's instructions for notation.

Point Reflection (*r*): If a point, A, and its image, A', are reflected under a point, P, then point P is the midpoint of the line connecting A and A'. $\triangle ABC$ is reflected under point P. $\triangle A'B'C'$ is its image.

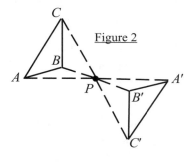

Figure 2

Symbols: $r_P(\triangle ABC) = \triangle A'B'C'$

or $\triangle ABC \xrightarrow{r_P} \triangle A'B'C'$

Properties: Distance, angle measure, parallelism, collinearity are all preserved under a point reflection. Orientation is the same in both figures - counterclockwise in this example. It is a direct isometry.

A figure can be reflected under a point that is also part of the figure. In the accompanying diagram, it shows a reflection on point C.

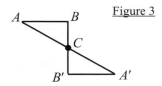

Figure 3

Translation (T): A shift in the figure in the "x" or horizontal direction and/or a shift in the "y" or vertical direction or both. Each point in the figure is translated the same distance and in the same direction. The value of the shift is added to the "x" and/or "y" coordinates of the point being translated to find the coordinates of its image. The rule can be written: $\triangle ABC \xrightarrow{T_{a,b}} \triangle A'B'C.'$ See also page 115, Translation of a Circle.

Example $\triangle ABC$ has vertices at $A(-5, 7)$, $B(-3, 4)$ and $C(-6, 1)$. After the translation, A' is located at $(4, 1)$. What are the coordinates of B' and C' ? This requires finding the translation rule and applying it to B and C.

$A(-5, 7) \rightarrow A'(4, 1)$

The "x" changed from -5 to $+4$, so 9 was added to "x".

The "y" changed from 7 to 1, so -6 was added to "y".

$B \xrightarrow{T9,-6} B'$

"x" coordinate: $-3 + 9 = 6$

"y" coordinate: $4 - 6 = -2$

B' is at $(6, -2)$

$C \xrightarrow{T9,-6} C'$

$x: -6 + 9 = 3$

$y: 1 - 6 = -5$

$C'(3, -5)$.

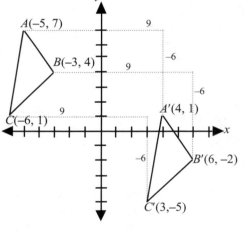

The rule could also be writen: $T_{9,-6} \triangle ABC \rightarrow \triangle A'B'C'$, which means take the x and y valuesof A, B, and C and add 9 to each x value and add (-6) to each y value to find the coordinates of $A'B'C'$, the image of ABC.

Properties: No change in distance, angles, parallelism, collinearity, or orientation. This is a direct isometry.

Rotation (R): The turning of a figure around a point, called the center of rotation, for a specified number of degrees. (A reflection over a point is the same as a rotation of 180°.) The center of rotation can be part of the figure or it may be outside the figure, but it does not move in the rotation. If no point is mentioned, then the rotation is assumed to be about the origin. Rotations are <u>counterclockwise</u> unless stated otherwise. A negative degree marking indicates a clockwise rotation.

Figure1

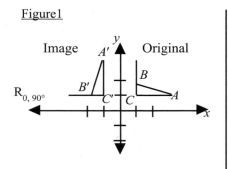

$R_{90°}$:*Counterclockwise* rotation about the origin

Figure 2

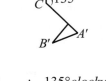

$R_{C,-135°}$: $-135°$*clockwise* rotation around point C. This is equivalent to a clockwise rotation of 225°.

Properties: Distance, angle measure, parallelism, collinearity are all perserved under a rotation. It is a direct isometry.

Dilation (D): The size of the image changes. A dilation is also called a similarity. The "x" and "y" coordinates of each point that is part of the dilation are multiplied by a constant to find the coordinates of its image. The center of dilation remains unchanged. The center of dilation is indicated in the notation if it is not the origin. A dilation of "k" where "k" is the constant of dilation, is a transformation such that:

   1. The image of point O, the center of dilation, is point O.

   2. The "x" and "y" coordinates of all other points are multiplied by "k" to find the coordinates of the images of the points.

Rule: Under a dilation of "k" whose center of dilation is the origin,
$P(x, y) \xrightarrow[D_k]{} P'(kx, ky)$ or $D_k(x, y) = (kx, ky)$.

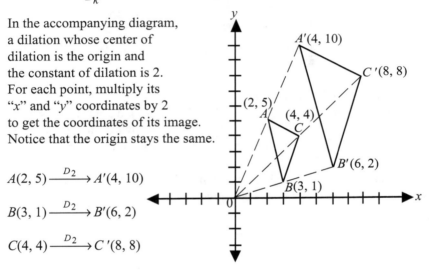

In the accompanying diagram, a dilation whose center of dilation is the origin and the constant of dilation is 2. For each point, multiply its "x" and "y" coordinates by 2 to get the coordinates of its image. Notice that the origin stays the same.

$A(2, 5) \xrightarrow{D_2} A'(4, 10)$

$B(3, 1) \xrightarrow{D_2} B'(6, 2)$

$C(4, 4) \xrightarrow{D_2} C'(8, 8)$

Note: If $0 \le k \le 1$ the image will be smaller than the preimage.
     If $k < 0$ the quadrant of the image will be different than the preimage.

Properties: DISTANCE is not preserved under a dilation. The image and its original or preimage are similar figures, but are not congruent (unless the constant of dilation is 1 in which case the figure is its own image). The other properties are still preserved: angle measure, collinearity, parallelism, and betweeness. It is not an isometry, it is a similarity.

Composite Dilations & Similarity: The image after a dilation is similar to the preimage - corresponding angles are equal and corresponding sides are proportional. A composition of dilations (see Composite Transformations on the next page) produces an image that is similar as well.

Geometry Made Easy

Composite Transformations: More than one tranformation can be performed on a figure. The image resulting from a line reflection can be rotated. A rotated image can be reflected or transformed. In a composition of transformations, it is necessary to perform the second transformation first, then perform the first transformation on that result.

Symbol (∘) – Example: $r_{y\text{-axis}} \circ r_{x\text{-axis}}$: means reflect over the x-axis, then reflect that over the y-axis

Example Given a triangle with vetices $A(2, 5)$, $B(3, 1)$, $C(7, 3)$. Find its image for $T_{-3, -5} \circ r_{y\text{-axis}}$.

Steps

1) First perform the reflection over the y–axis of the triangle.

Rule: $r_{y\text{-axis}} (x, y) \rightarrow (-x, y)$

$A(2, 5) \xrightarrow{\ r_{y\text{-axis}}\ } A'(-2, 5)$

$B(3, 1) \xrightarrow{\ r_{y\text{-axis}}\ } B'(-3, 1)$

$C(7, 3) \xrightarrow{\ r_{y\text{-axis}}\ } C'(-7, 3)$

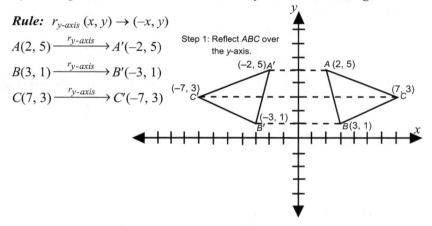

2) Then use those results to perform the translation

Rule: $T_{-3, -5}(x, y) \rightarrow (x - 3, y - 5)$

$A'(-2, 5) \xrightarrow{\ T_{-3, -5}\ } A''(-5, 0)$

$B'(-3, 1) \xrightarrow{\ T_{-3, -5}\ } B''(-6, -4)$

$C'(-7, 3) \xrightarrow{\ T_{-3, -5}\ } C''(-10, -2)$

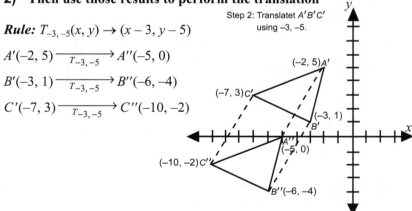

3) Conclusion: $\triangle ABC \xrightarrow{\ T_{-3, -5} \circ r_{y\text{-axis}}\ } \triangle A''B''C''$
$A''(-5, 0)$, $B''(-6, -4)$, $C''(-10, -2)$

<u>Glide Reflection:</u> A composite transformation in which a figure is reflected under a line and is translated along that line in a direction parallel to the line of reflection. In this particular composite transformation, the order of the transformations is not important. However, in order to avoid confusion, it is best to continue to work in the same order as we do for other composite transformations. Start at the right and work backwards.

Example Triangle ABC has vertices $A(2,-1)$, $B(-3,-2)$ and $C(-1,-3)$. Sketch and give the coordinates of the image of a glide reflection performed on ABC of $T_{-4,0}$ and r_{x-axis}.

Steps

1) Do the reflection first.

Rule: $r_{x-axis}\,(x, y) \rightarrow (x, -y)$

$A(2,-1) \xrightarrow{\ r_{x-axis}\ } A'(2, 1)$

$B(-3,-2) \xrightarrow{\ r_{x-axis}\ } B'(-3, 2)$

$C(-1,-3) \xrightarrow{\ r_{x-axis}\ } C'(-1, 3)$

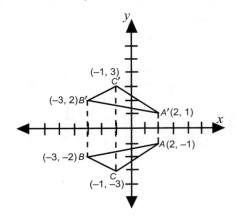

2) Then the translation

Rule: $T_{-4,0} \rightarrow (x-4, y)$

$A'(2, 1) \rightarrow A''(-2, 1)$

$B'(-3, 2) \rightarrow B''(-7, 2)$

$C'(-1, 3) \rightarrow C''(-5, 3)$

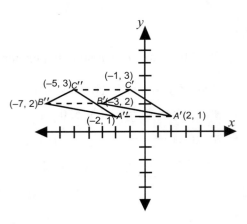

3) Conclusion: The coordinates of ABC after the glide reflection are $A''(0, 3)$, $B''(-3, 4)$, and $C''(-1, 5)$ $\triangle ABC \xrightarrow[\ T_{-4,0}\ \circ\ r_{x-axis}\ ]{} \triangle A''B''C''$

Note: Since the reflection and the translation are both isometries, the glide reflection is an isometry also.

11 – CONSTRUCTIONS

Construction: A drawing done in geometry using only a compass and a straight edge. Protractors, rulers, and graph paper or other devices for measuring are not permitted. No parts of a construction may be "sketched"- each part of a construction must be drawn with the compass and/or a straight edge. The lines and arcs used to make the construction are to be left on the paper. DO NOT ERASE the construction lines. In the constructions shown here the width of the compass opening should not be changed unless it is noted. The distance from the point of the compass to the point of the pencil serves as a measuring tool. In constructions, we are often duplicating the measure of something into a new drawing or we are using the compass to measure equal distances. If the compass should be changed from a previous step, it will be indicated. Also - the demonstration shown may not be the only way to do a construction. As we have seen, throughout this geometry book, there are often several ways to reach the same result. Provided that acceptable construction methods are used with valid logical reasoning, other "steps" may be used to do a construction.

Justify or explain the construction: This means to give a written sentence explaining why a particular method works. Sample justifications are given in the examples. Once again - follow your textbook and your teacher's instruction.

Examples Constructions

❶ Construct a line segment congruent to a given line segment.

Given: $\overline{AB}$

Construct: $\overline{CD} \cong \overline{AB}$

Steps:

1) Draw a line, "*m*", in a different location than $\overline{AB}$.

2) Put the compass point on A and the pencil point on B. This measures the length of $\overline{AB}$.

3) Move the compass point to any point on line "*m*" and label the point C

4) Swing an arc through line "*m*".

5) Label the point of intersection of "*m*" and the arc, D.

6) $\overline{AB} \cong \overline{CD}$

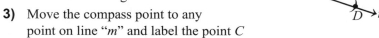

Justification: By using the compass to measure the length of the given segment, we can mark off an equal segment on another line.

❷ Construct the perpendicular bisector of a given line segment:

Given: $\overline{AB}$

Construct: $\overline{CD} \perp \overline{AB}$ and bisecting $\overline{AB}$

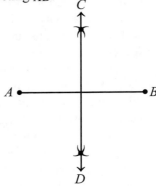

Steps::

1) Open the compass to more than 1/2 of $\overline{AB}$.

2) Put the compass point on A and swing two arcs - one above and one below $\overline{AB}$.

3) Move the compass point to B. Swing another pair of arcs, one above and one below $\overline{AB}$, so they intersect with the first pair.

4) Label the points where the arcs intersect, C and D.

5) Draw a line through point C and point D.

6) $\overline{CD}$ is the $\perp$ bisector of $\overline{AB}$.

Justification: All the points on the perpendicular bisector of a line segment are equidistant from the endpoints of the segment. Swinging 2 pair of equal intersecting arcs, one pair below $\overline{AB}$ and the other above $\overline{AB}$, intersecting at C and D, creates 2 points that are each equidistant from points A and B. The perpendicular bisector of $\overline{AB}$ goes through C and D.

❸ Construct a perpendicular line to a given line from a given point not on the line.

> Given: Line "*m*" and point *C* not on line "*m*".
> Construct: $\overline{CD} \perp m$.

Steps:

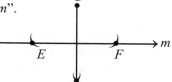

1) Put the compass point on *C*. Open the compass so the pencil point is on the other side of line "*m*".

2) Swing an arc below line "*m*" so it intersects line "*m*" in two places, *E* and *F*.

3) Put the point on *E* and swing an arc below "*m*". Then move the point to *F* and swing another arc below "*m*" so it intersects with the previous arc. Label the point of intersection of the two arcs, *D*.

4) Connect point *C* and point *D*.

5) $\overline{CD} \perp$ "*m*". ($\overline{CD}$ is actually the perpendicular bisector of the segment $\overline{EF}$ which we created and used on line "*m*".)

Justification: All the points on the perpendicular bisector of a line segment are equidistant from the endpoints of the segment. Creating *E* and *F* on line *m* that are equidistant from *C* makes two endpoints of a segment on line *m*. *D* is formed by swinging a pair of equal arcs from *E* and *F*. *D* is the second point that is equidistant from *E* and *F* and line $\overline{CD}$ is $\perp$ to line *m* through *C*.

❹ Construct a perpendicular line to a given line at a given point on the line.

Given: Line "*m*" and point *A* on line "*m*".

Construct: $\overline{AB} \perp m$ at *A*.

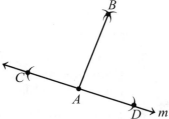

Steps:

1) Put the compass point on *A*. Swing an arc on each side of *A* through line "*m*". Label the points of intersection *C* and *D*.

2) Open the compass a little. Move the point to *C* and swing an arc above line "*m*". Move the point to *D* an swing another arc above line "*m*" so the two arcs intersect. Label the point of intersection of the two arcs, *B*.

3) Connect *B* and *A*.

4) $\overline{AB} \perp$ line "*m*" at *A*. (Here, too, we are making a segment on line "*m*" and then constructing the perpendicular bisector of that segment to get the perpendicular line required in the problem.)

Justification: All the points on the perpendicular bisector of a line segment are equidistant from the endpoints of the segment. Make endpoints, *C* and *D*, by swinging equal arcs from point *A* that intersect line *m*. Make another point, *B*, at the intersection of two equal arcs from points *C* and *D*. *A* and *B* are each equidistant from *C* and *D* so when *A* and *B* are connected a line perpendicular to line *m* at *A* is formed.

❺ Construct the bisector of a given angle.

Given: $\angle ABC$

Construct: $\overrightarrow{BD}$ so that $\overrightarrow{BD}$ bisects $\angle ABC$

Steps:

1) Put the compass point on B in the given angle. Swing an arc through each side of $\angle ABC$.

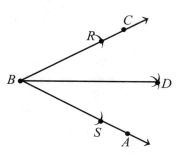

2) Label the points where the arcs intersect the sides of the angle with R and S.

3) Put the compass point on R and swing and arc - put it out in the area toward the opening of the angle.

4) Move the compass point to point S and repeat step 3.

5) Label the point of intersection of the two arcs D.

6) Draw a line from point B through point D.

7) $\overrightarrow{BD}$ is the bisector of $\angle ABC$.

Justification: All points on an angle bisector are equidistant from the sides of the angle. Locate two points R and S, one on each side of the angle, that are equally distant from the vertex. Swing a pair of equal intersecting arcs from each of those two points. The point of intersection, D, is equally distant from the sides the angle. Connect the intersection point with the vertex of the angle to form the angle bisector.

❻ Construct an angle congruent to a given angle.

Given: ∠ABC

Construct: ∠DEF ≅ ∠ABC

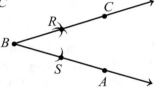

Steps:

1) Draw a line "*m*" and place a point, *E* on the line.

2) Put the compass point on *B* in the given angle. Swing an arc through each side of ∠ABC. Label the points where the arcs intersect the sides of the angle with *R* and *S*.

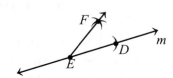

3) Move the compass point to point *E* on line "*m*".

4) Swing an arc that intersects "*m*" and label the point of intersection *D*.

5) With the point still at *E*, swing another arc in approximately the place that the other side of the angle will be.

6) Put the compass point on *R* and the pencil point on *S* to measure that distance.

7) Move the point of the compass to point *D*.

8) Swing an arc from point *D* to intersect with the arc from step 5. Label that point *F*.

9) Connect point *F* and point *E*.

10) ∠DEF ≅ ∠ABC

Justification: If two angles are congruent, the measure between their sides at any specific point is equal. By marking off two equal segments on the sides of the given angle and measuring between the endpoints of the segments we can determine the measures of corresponding parts of the congruent angle.

Constructions 7-8 are based on knowledge of the first 6 constructions

❼ Construct a line parallel to a given line through a given point not on the line.

Given: Line $\overleftrightarrow{MN}$ and Point A not on $\overleftrightarrow{MN}$.
Construct: Line $S \parallel \overleftrightarrow{MN}$.

Plan: Use the knowledge that if two lines are cut by a transversal making a pair of corresponding angles that are congruent, the lines are parallel.

Steps:

1) Draw a line through point A to intersect with $\overleftrightarrow{MN}$ at point F creating $\angle MFA$. ($\angle 1$)(This line, $\overleftrightarrow{AF}$, will become the transversal.)

2) Using point A as the vertex of the angle and line $\overleftrightarrow{AF}$ as one side of the angle, construct $\angle 2$, congruent to $\angle 1$ with its vertex at A to make a pair of congruent corresponding angles. $\angle 1 \cong \angle 2$ (See construction 6)

3) The new side of the congruent angle is the line parallel to $\overleftrightarrow{MN}$. Extend that side and label it S.

4) $S \parallel \overleftrightarrow{MN}$

Justification: Two parallel lines have corresponding angles that are congruent. By drawing a transversal through the line and also through the given point, the vertex and one side of the corresponding angle are formed. The second side of a congruent angle on the transversal at the given point results in a parallel line through the vertex being created.

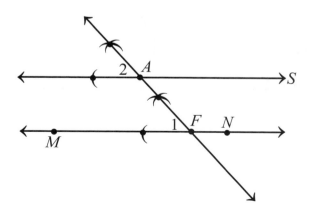

❽ Construct an equilateral triangle.

Given: A segment congruent to each of the three congruent sides of the triangle.

Steps:

1) Construct a line segment equal to the given segment as shown in construction #1.

2) Keeping the compass unchanged, create an arc from each endpoint of the new segment so the arcs intersect.

3) The point of intersection is the vertex of the equilateral triangle. Connect it to each endpoint of the segment.

Justification: In an equilateral triangle, all three sides are congruent. Copy the length of the given segment with the compass and mark off a congruent segment for one side of the triangle. Swinging two more equal arcs gives a point of intersection that is the same distance from each endpoint. Connect the point of intersection to each endpoint to form two more congruent sides.

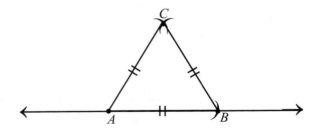

12 — LOCUS OF POINTS

<u>Locus</u>: The set of all points, and only those points, which satisfy a given condition or conditions. (The plural is loci.) In this section, all loci will be in a plane. The terminology used is "a locus of points..." to describe the set of points required. Two or more loci can be combined to locate the points common to both (or all) the described loci. The locus is demonstrated with dotted lines.

To find a locus it is not always required that graph paper be used, but it often helps as the distances can be counted and lines drawn with ease.

Steps:

1) Make a diagram (or graph) of the lines or points that are given.

2) Analyze the condition that must be satisfied.

3) Locate and mark on the graph a point which satisfies the given condition. Then locate several more points that also satisfy the conditions. Be careful to cover all the possible conditions - on both sides of a line, all the way around a point, etc.

4) Use the points to draw a line or a smooth curve to show the anticipated locus. (A dotted line is often used to show the locus.)

5) Check several other points on the locus you have drawn to make sure that your estimate of where the locus is located is correct.

6) Write in words a description of the figure that is the locus.

<u>Compound Loci</u>: Combinations of loci. Do steps 1–5 above for each condition given. Then describe in words the figure or points that result from the intersection of the loci. (See page 99 for examples.)

Basic Loci: Some loci are shown with examples here, some are general, and some are just diagrams and descriptions. You must be able to recognize and use them all.

The locus of points a given distance from a fixed point:

The locus of points a given distance from a fixed point is a circle with the given point as its center and a radius equal to the given distance.

Sketch point A at the origin. Locate C, D, E, and F by counting three units along each axis away from A. Sketch points G, H, I, and J to estimate where the locus will be. Use all those points to draw the locus.

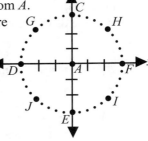

Conclude: The locus of points three units from A is a circle with its center at A and a radius of 3.

The locus of points equidistant between two fixed points

The locus of points equidistant between two fixed points, A and B is a line which is the perpendicular bisector of a segment, $\overline{AB}$, joining the two points.

Sketch the two points on a graph. Connect A and B. Then locate several points that are equally distant from them both. Connect the new points to sketch the locus.

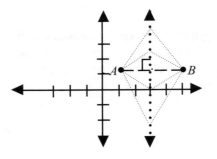

Conclude: The locus of points equidistant from A and B is a line which is the $\perp$ bisector of the line segment $\overline{AB}$.

The locus of points equidistant from two intersecting lines

The locus of points equidistant from two intersecting lines, $\overleftrightarrow{AB}$ and $\overleftrightarrow{CD}$ which intersect at E, is a pair of lines which bisect the angles formed by the two intersecting lines.

(This one is difficult to sketch and measure unless a compass is used.)

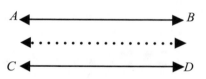

The locus of points equidistant from two parallel lines

The locus of points equidistant from two parallel lines, $\overleftrightarrow{AB}$ and $\overleftrightarrow{CD}$, ($\overleftrightarrow{AB} \parallel \overleftrightarrow{CD}$) is a line parallel to both lines and midway between them.

The locus of points equidistant from a line

The locus of points equidistant from a line, $\overleftrightarrow{AB}$, is a pair of lines, $\overleftrightarrow{CD}$ and $\overleftrightarrow{EF}$, one on each side of $\overleftrightarrow{AB}$, each is parallel to $\overleftrightarrow{AB}$, and each is the given distance from $\overleftrightarrow{AB}$.

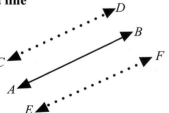

Writing the equation for a locus of points: Use graph paper to sketch the problem to see if the locus is actually represented by the equation that you find.

Example Write an equation of the locus of points equidistant from $y = 2x + 2$ and $y = 2x - 4$.

To solve: The two given lines both have a slope = 2. They are parallel. One has a y-intercept of –4, the other intercepts the y-axis at 2. The locus of points equidistant from two parallel lines is a line parallel to them and midway between them.

The locus also has a slope = 2 since it is parallel to the other lines. The "y" intercept of the locus is halfway between the intercepts of the two lines, –4 and +2. It is at $(0, -1)$. The equation for the locus is $y = 2x - 1$.

Example Write an equation of the locus of points two units from $(-3, 4)$.

Analysis: This is a circle with the center at $(-3, 4)$. Use the standard formula for a circle, $(x - h)^2 + (y - k)^2 = r^2$. The center of the circle is (h, k) is given here at $(-3, 4)$ and r is given as 2. The equation is $[x - (-3)]^2 + (y - 4)^2 = 2^2$ which may be left in the form: $(x + 3)^2 + (y - 4)^2 = 4$

Example Write an equation of the locus of points equidistant from the points $A(-5, 0)$ and $B(3, 4)$.

Analysis: From the basic loci, we know that this is a line that is the perpendicular bisector of the line segment joining the two points.

To solve: Sketch the points and segment, $\overline{AB}$, that join them. Find the slope of $\overline{AB}$. Find the midpoint of $\overline{AB}$. The perpendicular bisector of $\overline{AB}$ will go through the midpoint and have a slope that is the negative reciprocal of $\overline{AB}$.

Slope of $\overline{AB}$: $m_{\overline{AB}} = \dfrac{4 - 0}{3 - (-5)} = \dfrac{4}{8} = \dfrac{1}{2}$

Therefore, the slope of the locus is –2.

Midpoint $\overline{AB}$: $m_{\overline{AB}} = \left(\dfrac{-5 + 3}{2}, \dfrac{0 + 4}{2} \right) = (-1, 2)$

Therefore, the locus goes through $(-1, 2)$

Use the point slope formula to find the equation of the locus

$$y - 2 = -2(x - (-1))$$
$$y - 2 = -2(x + 1)$$
$$y - 2 = -2x - 2$$
$$\underline{ +2 +2 }$$

The equation for the locus is $\qquad\qquad y = -2x$

Geometry Made Easy

Compound Loci: Draw the first locus on a graph. Then draw the second locus on the same graph. The points where the two loci intersect is the location of the compound loci because they satisfy both conditions. USE GRAPH PAPER for these problems - it is easy to sketch a locus inaccurately on plain paper and the intersection of the loci may not be right.

Example Find the locus of points 2 units from the line $x = 4$ and 5 units from the point (3, 2).

Solution: Draw the line $x = 4$ on the graph. Draw the 2 parallel lines that are the locus of points 2 units from $x = 4$. Then draw the circle with its center at (3, 2) and a radius of 5. The point(s) of intersection of the loci A, B, C, and D are the solution.

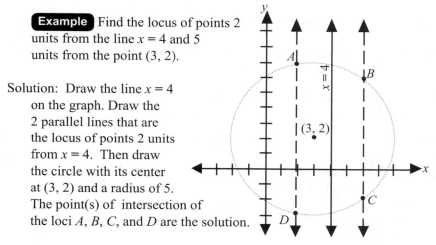

In the example above, there are 4 points since the circle with a radius of 5 intersects both lines. There would be 3 points if the circle intersected one line and was tangent to the other, and only 2 points if the circle had a radius small enough that it only intersected one of the parallel locus lines, or even one point if the circle was tangent on one parallel locus line. Accurate sketching of the graph is important.

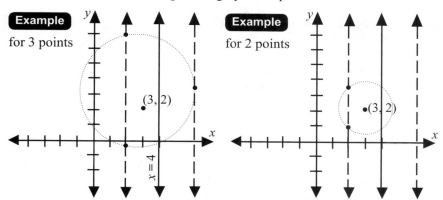

Example for 3 points

Example for 2 points

Special Locus: The locus of points equidistant from a point and a fixed line is a parabola. The fixed point is called the **focus** and the fixed line is the **directrix**.

13 — CIRCLES

Circle: The locus of points a given distance (called the radius) from a point called the center. A circle is measured in degrees. There are 360° in a circle. A circle is named by its center point.

Radius: A line segment from any point on the circle to the center.

Diameter: A line segment from any point on the circle that goes through the center to another point on the circle.

Circumference: The distance around the edge of the circle.
  Formula: $C = 2\pi r$ or $C = \pi d$

Area: The space enclosed by or inside the circle. Formula: $A = \pi r^2$

Subtend: When the end points of an arc are formed by the intersection of the rays of an angle and the circle, the $\angle$ subtends (or cuts off) the arc.
$\angle A$ subtends $\overset{\frown}{CD}$.

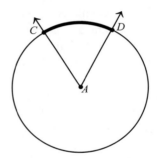

Measure of a Circle: A circle has a measure of 360°.

Arc: Part of the circumference. Arc CD is written $\overset{\frown}{CD}$. It means the part of the circumference that is between points C and D. It can be measured in degrees and it can be measured in linear measure called the arc length.

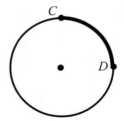

Locating the Arc: Any arc can be labeled with multiple points – depending on what points are labeled on the circle between the endpoints of the arc. When locating the arc, start on the first letter named and move toward the next letter named in the shortest way. Continue until you get to the last letter named.

Geometry Made Easy

Types of Arcs:

1. Minor Arc: An arc that measures < 180°.
   It is labeled with 2 letters. $\overset{\frown}{BC}$ in figure 1.

2. Major Arc: An arc that measures > 180°.
   It is usually labeled with 3 letters. $\overset{\frown}{CDF}$ in figure 1.

3. Semi-circle: An arc that measures 180°. Its points on the circle are the endpoints of a diameter of the circle. $\overset{\frown}{FCD}$ in figure 1.

Given: Circle A

Radius: $\overline{AC}$

Diameter: $\overline{DF}$

Arcs: Can be labeled in either direction. $\overset{\frown}{BC}$ and $\overset{\frown}{CB}$ are the same arc. These are some of the labels:

Figure 1

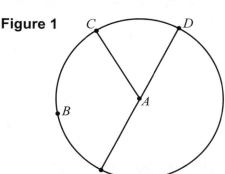

1. Minor Arcs: $\overset{\frown}{BC}$, $\overset{\frown}{BD}$ or $\overset{\frown}{BCD}$, $\overset{\frown}{CD}$, $\overset{\frown}{BF}$, $\overset{\frown}{CBF}$ or $\overset{\frown}{CF}$ are all minor arcs. They each measure < 180°.

2. Major Arcs: $\overset{\frown}{BCF}$ or $\overset{\frown}{BCDF}$, $\overset{\frown}{CDF}$, $\overset{\frown}{CDFB}$ or $\overset{\frown}{CFB}$ are all major arcs. Each measures > 180°.

3. Semi Circles: $\overset{\frown}{DF}$ or $\overset{\frown}{FD}$. These are two ways to name the semicircle. Each measures 180°.

Chords, Tangents and Secants

Chords, tangents, and secants are the types of lines and segments that are used with circles. (See also pages 107-110)

Chord: A segment that connects two points on a circle. It is in the interior of the circle.

Tangent: A line that extends from a point outside the circle to the circle, touching the circle at exactly one point. Tangent $\overleftrightarrow{IJ}$ is tangent to circle A at M in figure 2 on the next page. It does not enter the circle. If a tangent is extended past the circle, its extension goes "past" the circle, not into it. We often use part of a tangent as a segment. ($\overline{IM}$ or $\overline{JM}$ in figure 2 on the next page)

Secant: A line that passes through two points on a circle. We often use part of a secant as a segment. It connects a point outside the circle, intersects the circle and goes across it to another point on the other side of the circle. (see $\overline{LR}$ in figure 2)

Chord: $\overline{BH}$ Another chord is $\overline{RQ}$ which is part of the secant $\overline{RL}$.

Tangent: $\overline{IJ}$ intersects circle A at M.

Secant: $\overline{RL}$ which intersects circle A at R and Q.

Figure 2

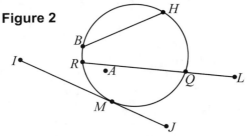

Circles and their Angles and Arcs

The angles formed by the segments and lines associated with a circle have rules for finding their measure that are related to the arcs they intercept or subtend. Some have special names, others are named descriptively. If the measures of the arcs are available, the measures of the related angles can be found. And conversely, if the measures of the angles are available, the measures of the arcs can be found.

Sum of the Arcs

The sum of the arcs in any circle is 360°. Ratios are sometimes used to find the degree measure of the arcs in a circle.

Examples

❶ In circle O, the ratio of $\overarc{ABC}$ to $\overarc{AC}$ is 3:2. Find the degree measure of each arc.

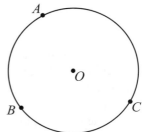

Let x be the ratio factor.
Therefore $\overarc{ABC} = 3x$ and $\overarc{AC} = 2x$
$3x + 2x = 360$
$.5x = 360$
$x = 72$
$\overarc{ABC} = 3(72) = 216°$
$\overarc{AC} = 2(72) = 144°$

Geometry Made Easy

❷ Arcs $\overset{\frown}{ABC}$, $\overset{\frown}{CD}$, $\overset{\frown}{DA}$ and are in a ratio of 5:2:3 respectively. Find the measure of each arc.

Let $5x = \overset{\frown}{ABC}$

Therefore $2x = \overset{\frown}{CD}$, $3x = \overset{\frown}{DA}$

$2x + 3x + 5x = 360$

$10x = 360$

$x = 36$

$\overset{\frown}{ABC} = 5(36) = 180°$

$\overset{\frown}{CD} = 2(36) = 72°$

$\overset{\frown}{DA} = 3(36) = 108°$

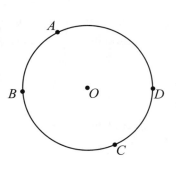

Internal Angles

Central angle: Vertex is at the center of the circle, both sides are radii. The intercepted arc = the degree measure of the central angle.

Inscribed angle: The vertex is on the circle and the sides are chords (figure 3), a chord and a secant (figure 4), or 2 secants. The measure of the inscribed angle $= \frac{1}{2}$ the measure of the arc it intercepts. See Figure 3 and Figure 4.

Given: Circle O **Figure 3**

radii $\overline{QO}$ and $\overline{OP}$

chords $\overline{ST}$ and $\overline{SR}$

Arc measures as labeled

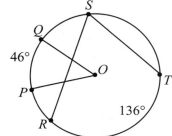

Examples

❶ **Central Angle:** $m\angle QOP = m\overset{\frown}{QP}$; $m\angle QOP = 46$; $m\overset{\frown}{QP} = 46$

> **Note:** A central angle is the <u>only</u> angle in the group of angles related to circles that is equal to the arc it cuts off.

> **Note:** A diameter forms a central angle of 180° since it intercepts $\frac{1}{2}$ of the 360° circle. See Figure 3

❷ **Inscribed angle:** $m\angle TSR = \frac{1}{2}\,\overset{\frown}{TR}$; $m\angle TSR = \frac{1}{2}(136) = 68$

Other angles with a vertex on the circle are not inscribed angles. These angles are formed by a tangent and a secant ($\angle 1$ in Figure 4) or a tangent and a chord ($\angle 2$ in Figure 4): Vertex is on the circle, one side is a tangent and the other is a chord or secant. The measure of this angle $= \frac{1}{2}$ its intercepted arc.

Figure 4

Given: Circle A

Tangent $\overline{CED}$

Chord $\overline{EF}$

Secant $\overline{FG}$

Arc measures as labeled

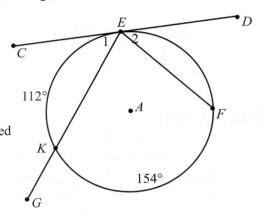

$m\angle 1 = \frac{1}{2} m \overset{\frown}{EK}$; $m\angle 1 = \frac{1}{2}(112) = 56$

$m\angle 2 = \frac{1}{2} m \overset{\frown}{EF}$; $m\angle 2 = \frac{1}{2}(94) = 47$

Find $m \overset{\frown}{EF}$ first: $360 - (112 + 154) = 94$

Inscribed $m\angle KEF = \frac{1}{2} \ m\overset{\frown}{KF}$; $m\angle KEF = \frac{1}{2}(154) = 77$

- *No matter whether the sides are tangents, chords, or secants, the same rule applies. When the vertex is ON the circle, the angle measures $\frac{1}{2}$ the measure of its intercepted arc.*

External Angles

The vertex is away from the circle. It is formed by tangents and secants: The measure of the angle $= \dfrac{1}{2}$ the difference of the arcs it intercepts.

1. External angle formed by 2 tangents: The vertex is outside the circle. Both sides are tangents. $\angle BCE$ in Figure 5. See Example ❶
2. External angle formed by a secant and a tangent: $\angle BCD$ and $\angle DCE$ in Figure 5. See Example ❷
3. Exterior angle formed by 2 secants. $\angle WFD$ in Figure 5. See Example ❸

Examples

❶ $m\angle \text{BCE} = \dfrac{1}{2}(m\,\widehat{BWE} - m\,\widehat{BJE})$

$m\,\widehat{BWE} = 45 + 24 + 52 + 82 + 52 = 255$

$m\,\widehat{BJE} = 75 + 30 = 105$

$m\angle \text{BCE} = \dfrac{1}{2}(255 - 105) = 75$

Figure 5

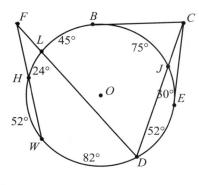

❷ Two different examples are shown here.

1. $m\angle BCD = \dfrac{1}{2}(m\widehat{BWD} - m\widehat{BJ})$

$m\angle BCD = \dfrac{1}{2}(203 - 75) = 64$

2. $m\angle DCE = \dfrac{1}{2}(m\widehat{DE} - m\widehat{JE})$

$m\angle DCE = \dfrac{1}{2}(52 - 30) = 11$

❸ $m\angle WFD = \dfrac{1}{2}(m\widehat{WD} - m\widehat{HL})$ $m\angle WFD = \dfrac{1}{2}(82 - 24) = 29$

Note: Each of this group of angles has its vertex outside or "away" from the circle. If you remember that "take <u>away</u>" means subtract, it might help you remember that you have to subtract the smaller arc from the larger one and then divide by 2 for each of these angles.

- **When the vertex is external (away from the circle), its measure is $\dfrac{1}{2}$ the difference of its intercepted arcs.**

Interior Angles Formed By Two Chords

Two chords that intersect in a circle: An angle formed by two chords intersecting inside a circle is equal to ½ the sum of arcs it intercepts and the arc its vertical angle intercepts.

Given: Circle A,

chords $\overline{FE}$ and $\overline{CD}$.

Arc measures as labeled.

Figure 6

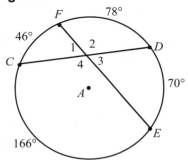

$\angle 1$ and $\angle 3$ are vertical angles.

$m\angle 1 = \dfrac{1}{2}(m\,\overset{\frown}{CF} + m\,\overset{\frown}{DE})$

$m\angle 1 = \dfrac{1}{2}(46 + 70) = 58$

$m\angle 3 = 58$

2 and 4 are vertical angles.

$m\angle 2 = \dfrac{1}{2}(m\,\overset{\frown}{FD} + m\,\overset{\frown}{CE})$

$m\angle 2 = \dfrac{1}{2}(78 + 166) = 122$

$m\angle 4 = 122$

Note: I help my students remember this is a sum because the intersection of the two chords inside the circle looks something like a plus sign. This is the only type of angle we study here that is equal to $\dfrac{1}{2}$ the sum of the arcs intercepted by the angle and its vertical angle.

This is the only type of angle that $= \dfrac{1}{2}$ the SUM of its arc and the arc of its vertical angle.

Circles and Segments

Finding the length of various segments of chords, secants, and tangents that intersect each other in a given circle requires that different formulas be used to calculate the lengths. These problems sometimes result in a quadratic equation that must be solved.

- Two chords that intersect in a given circle: The product of the segments of one chord is equal to the product of the segments of the other.

  In circle A, chords $\overline{BE}$ and $\overline{CD}$ intersect at F. (See figure 7)
  Rule: $(BF)(FE) = (CF)(FD)$

Examples

❶ If $BF = 5$, $CF = 7$, and $FD = 6$, find FE.
Let $x = FE$
$5x = 7(6)$
$x = 8\dfrac{2}{5}$ or 8.4
$FE = 8.4$

Figure 7

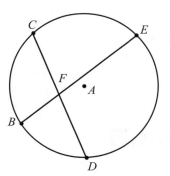

❷ Given: $CF = FD$, $FB = 2$,
$FE = 12$ more than CF.
Find the length of FE.
Let $x = CF$ and FD
$\therefore x + 12 = FE$
$(x)(x) = 2(x + 12)$
$x^2 = 2x + 24$
$x^2 - 2x - 24 = 0$
$(x - 6)(x + 4) = 0$
$x = 6$, $x = -4$ reject
$FE = x + 12 = 6 + 12 = 18$

- Two tangents that share one external point: When two tangents share the same external point, the segments from that point to the points of tangency on the circle are equal.

Figure 8

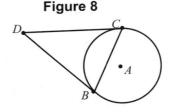

Given: Circle A with tangents $\overline{DB}$ and $\overline{DC}$ that intersect at D. Chord BC.

Prove: $\triangle BDC$ is isosceles.

Solution:
Circle Proof in paragraph form: Since we are given two tangents, $\overline{DB}$ and $\overline{DC}$, that intersect at D, we know that $\overline{DB} \cong \overline{DC}$ because two tangents that share the same external point are congruent (or equal). That makes two sides of $\triangle BDC$ congruent. Therefore, $\triangle BDC$ is isosceles because the definition of an isosceles triangle is that two sides must be congruent.

- Two secants that intersect at the same external point: When two secants are drawn from the same external point to a circle; the product of the external segment of one secant and its entire length is equal to the product of the external segment of the other and its entire length. In the example below, $(DE)(DC) = (DF)(DB)$.

Figure 9

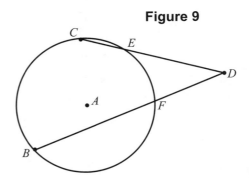

If $CE = 2$, $DE = 4$,
and $BF = 5$, find DF.
Let $x = DF$
$(4)(4 + 2) = x(x + 5)$
$24 = x^2 + 5x$
$x^2 + 5x - 24 = 0$
$(x + 8)(x - 3) = 0$
$x = -8$ reject, $x = 3$
$DF = 3$

- A tangent and a secant intersect at the same external point: The length of the tangent is the mean proportional between the external segment of the secant and the entire secant. The same situation can be described this way: The tangent squared is equal to the product of the external segment of the secant and the entire secant.

Example $\dfrac{DE}{CD} = \dfrac{CD}{BD}$ or $(CD)^2 = (BD)(DE)$

If $BE = 9$ and $DE = 3$, find CD.
two solutions are shown

Let $x = CD$

Figure 10

Method 1	Method 2
$\dfrac{3}{x} = \dfrac{x}{(9+3)}$	$x^2 = (3)(9+3)$
$x^2 = 36$	$x^2 = 36$
$x = \pm 6$, reject -6	$x = 6$
$CD = 6$	$CD = 6$

Chords, tangents, radii, diameters, and arcs

These are additional statements that can be used in solving circles or doing circle proofs. "Solving" a circle means to find as much information as possible about its arc, angles, and segments.

1. If a radius is perpendicular to a chord, it bisects the chord and its arc.

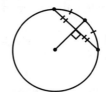

2. If a diameter or radius meets a tangent at the point of tangency, it is perpendicular to the tangent.

3. If two chords are equal they intercept equal minor arcs.

4. If two minor arcs are =, the corresponding chords are ≅.

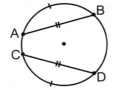

5. If two chords or secants in a circle are parallel, the arcs between them are equal.

$$\overline{AB} \parallel \overrightarrow{CD}$$

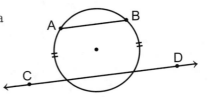

6. In a circle or in congruent circles, if two chords are equidistant from the center, they are congruent.

Given: $\odot H \cong \odot R$

$\overline{HT} \cong \overline{HS} \cong \overline{RU}$

$\therefore \overline{AB} \cong \overline{CD}$

Since $\odot H \cong \odot R$, and

$\overline{RU} \cong \overline{HS}$, $\overline{LE} \cong \overline{AB}$

and $\overline{LE} \cong \overline{CD}$.

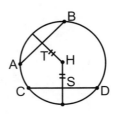

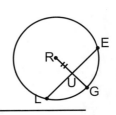

Common tangents of two circles: A tangent line can be tangent to two circles at the same time in various situations depending on the location of the circles in relation to each other. It is called a "common tangent." The diagrams below show the possible configurations of circles and common tangents. Even though a tangent is associated with two circles, all the same rules apply relative to each circle for tangent segment length, angle measure, perpendicularity to an intersecting radius or diameter, etc.

Two circles that are completely separate have four common tangents.

The tangents are numbered and the points of tangency are labeled with letters.

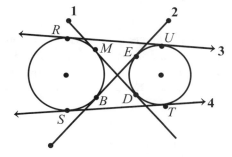

Two circles that are *tangent to each other externally* have three common tangents. Tangent externally means the circles share one common point and they are not inside each other.

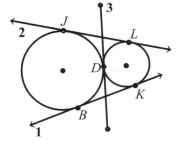

Two circles that intersect each other in two points have two common tangents.

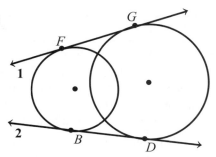

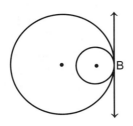 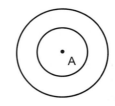 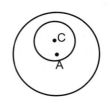

Two circles that are internally tangent have one common tangent line.

Concentric circles have no common tangents. They share the same center.

Circles that are inside each other but do not touch have no common tangents.

Circle Proof – statement reason form

Given: Circle E with diameters $\overline{DB}$ and $\overline{CA}$

Prove : $\triangle CDE \cong \triangle ABE$

Statement	Reason
1. Circle E with diameters $\overline{BD}$ and $\overline{CA}$	1. Given
2. $\angle CED \cong \angle AEB$	2. Vertical angles are congruent
3. $\overarc{CD} \cong \overarc{AB}$	3. In a circle, two arcs intercepted by congruent central angles are congruent.
4. $\overline{CD} \cong \overline{AB}$	4. In a circle, if two minor arcs are congruent, their corresponding chords are congruent.
5. $\angle CDB \cong \angle CAB$	5. If two inscribed angles intercept the same or equal arcs, they are congruent.
6. $\triangle CDE \cong \triangle ABE$	6. $AAS \cong AAS$

Note: This proof could be done several other ways

- Segments EB, EA, CE, and DE are all radii and therefore congruent. Since $\angle CED$ and $\angle BEA$ are vertical $\angle$'s, they are congruent. That makes arcs $\overarc{CD}$ and $\overarc{AB}$ congruent. In turn, that makes chords $\overline{CD}$ and $\overline{AB}$ congruent. The triangles are congruent by $SSS \cong SSS$.

- Perhaps the easiest way to prove these two triangles are congruent would be to use SAS. The radii are all equal and $\angle CED \cong \angle BEA$. The triangles are congruent by $SAS \cong SAS$. Any of these three methods is perfectly acceptable.

Geometry Made Easy

Solve: In circle A, diameter $\overline{DE}$, secant $\overline{HG}$, and tangent $\overline{FG}$ are drawn. $\overline{DE}$ intersects $\overleftrightarrow{FG}$ at E. The ratio of the arcs $\overset{\frown}{HD} : \overset{\frown}{DB} : \overset{\frown}{BE}$ is 2:7:5. Find the measure of each of the numbered angles.

Solution: Let x be the ratio factor.

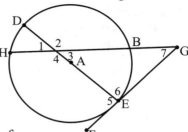

$\therefore m\overset{\frown}{HD} = 2x$; $m\overset{\frown}{DB} = 7x$; and $m\overset{\frown}{BE} = 5x$. Since $\overline{DE}$ is a diameter, $\overset{\frown}{DBE} = \overset{\frown}{DB} + \overset{\frown}{BE} = 180°$, also $\overset{\frown}{DHE} = \overset{\frown}{DH} + \overset{\frown}{HE} = 180°$

Make an algebraic equation and solve for x.

Find x

$\overset{\frown}{DB} + \overset{\frown}{BE} = 180°$

$7x + 5x = 180°$

$12x = 180°$

$x = 15°$

Find all the arc measures

$\overset{\frown}{DB} = 7(15) = 105°$

$\overset{\frown}{BE} = 5(15) = 75°$

$\overset{\frown}{DH} = 2(15) = 30°$

$\overset{\frown}{EH} = 180 - 30 = 150°$

Now work on the angles. It is very helpful to label the measure of each arc on your diagram. There are three different kinds of angles involved. Show the calculations.

1) Angles 1, 2, 3 and 4 are vertical angles formed by the intersection of a chord and a secant.

   $m\angle 1 = (\frac{1}{2})(30 + 75) = 52.5$

   $m\angle 3 = m\angle 1 = 52.5$

   $m\angle 2 = (\frac{1}{2})(105 + 150) = 127.5$

   $m\angle 4 = m\angle 2 = 127.5$

2) Angles 5 and 6 have vertices ON the circle. Since they are formed by a diameter intersecting with a tangent, both are right angles. Or you can use the formula $(\frac{1}{2})$ (arc intercepted)

   $m\angle 5 = (\frac{1}{2})(180) = 90$

   $m\angle 6 = (\frac{1}{2})(105 + 75) = 90$

3) Angle 7 is formed by a tangent and secant drawn from the same external point. It is an external angle and $= (\frac{1}{2})$(difference of the arcs it intercepts.) $m\angle 7 = (\frac{1}{2})(150 - 75) = 37.5$

Circles and Graphs

Equation of a circle: $(x - h)^2 + (y - k)^2 = r^2$ where (h, k) are the coordinates of the center of the circle and "r" is the radius.

Examples

❶ Write the equation of a circle with its center at (4, −3) and a radius of 6.

Solution: Substitute the coordinates of the center, (4, −3), for "h" and "k" in the formula and substitute 6 for "r".

$(x - 4)^2 + (y - (-3))^2 = 6^2$ which becomes $(x - 4)^2 + (y + 3)^2 = 36$

The circle equation is often left in the form shown but it can be multiplied out if necessary.

❷ Graph the following: $(x - 2)^2 + (y + 2)^2 = 25$

Solution: Determine the coordinates of the center (h, k) by comparing the equation with the known formula. "h" is 2 and "k" is −2. So the center is at (2, −2). Plot that on the graph and label it as the center. Since 25 is the radius squared, then $r = 5$. Using the grid, place points on the graph 5 units vertically and horizontally from (2, −2). Then estimate a few other points and sketch in the circle.

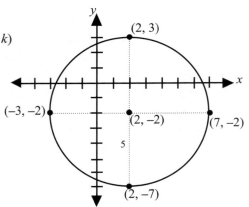

Transformations on circles: Find the coordinates of the center and of a point on the preimage. Use those points to calculate the coordinates of the corresponding points on the image.

Translation of a circle: Change the coordinates of the center but keep the radius unchanged to demonstrate a translation.

Example Graph both circles and write an equation for circle A after completing the following transformation.

$\odot A : (x - 2)^2 + (y - 3)^2 = 16$

$\odot A \xrightarrow{T_{3, -5}} \odot A'$

Solution: The center of $\odot A$ is (2, 3). The radius is 4. Graph Circle A. Add 3 to the "x" value of the center of A and -5 to the "y" value. The center of A' is at $(2 + 3, 3 - 5)$ or $(5, -2)$. Graph A' with a radius of 4.

Equation of A': $(x - 5)^2 + (y + 2)^2 = 16$

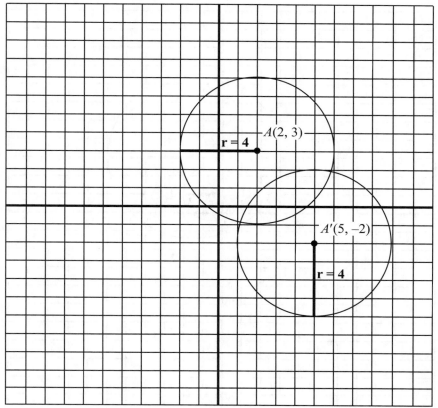

Dilation of a circle:

To perform a dilation on a circle with the origin as the center of dilation, multiply the coordinates of the center, the coordinates of a point on the circle, and the radius by the dilation factor.

Example Graph the circle A which is represented by the equation $(x - 1)^2 + (y - 2)^2 = 9$. Circle A is transformed by the following rule: $A \xrightarrow{D_3} A'$. The center of dilation is the origin. Graph A', state the coordinates of the center, a point on the circle, and the length of the radius of A'.

Write the equation representing the circle A'

Solution: Circle A has a center at $(1, 2)$ and the radius is 3. A point on the circle is $(4, 2)$. Circle A' has a center at $(3, 6)$ and the radius is 9. A point on A' is $(12, 6)$.

The equation of A' is $(x - 3)^2 + (y - 6)^2 = 81$

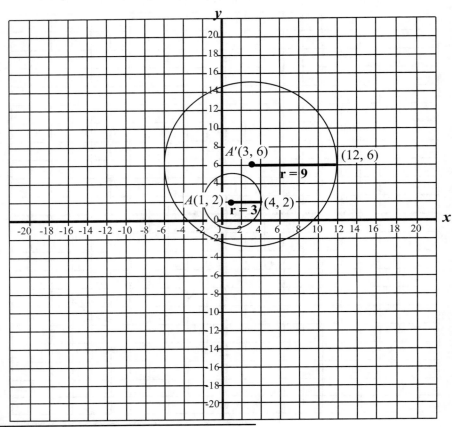

New York State Mathematics Glossary

This Glossary provides an understanding of the mathematical terms used in the Regents-approved course entitled Geometry as reflected in the New York State Mathematics Core Curriculum (Revised 2005). We encourage all students to become familiar with these terms .

A

AA triangle similarity: If there exists a one-to-one correspondence between the vertices of two triangles such that two angles of one triangle are congruent to the corresponding two angles of the second triangle, then the two triangles are similar.

AAS triangle congruence: If there exists a one-to-one correspondence between the vertices of two triangles such that two angles and the side opposite one of them in one triangle are congruent to the corresponding parts of the second triangle, then two triangles are congruent.

ASA triangle congruence: If there exists a one-to-one correspondence between the vertices of two triangles such that two angles and the included side of one triangle are congruent to the corresponding parts of the second triangle, then two triangles are congruent.

abscissa: The horizontal or x-coordinate of a two-dimensional coordinate system.

absolute value: The distance from 0 to a number n on a number line. The absolute value of a number n is indicated by $|n|$.

Example: $|-3| = 3$, $|+3| = 3$, and $|0| = 0$

acute angle: An angle whose measure is greater than $0°$ and less than $90°$.

acute triangle: A triangle that contains three acute angles.

additive property of equality: If a, b and c are real numbers such that $a = b$, then $a + c = b + c$.

adjacent angles: Two coplanar angles that share a common vertex and a common side but have no common interior points.

adjacent sides: Two sides of any polygon that share a common vertex.

algebraic representation: The use of an equation or algebraic expression to model a mathematical relationship.

algorithm: a defined series of steps for carrying out a computation or process.

alternate interior angles: Any two nonadjacent angles that lie on opposite sides of a transversal and that are interior to the lines.

altitude –

    Of a cone: A line segment drawn from the vertex of the cone perpendicular to the plane containing its base.

    Of a cylinder: A line segment drawn from any point on one base of a cylinder perpendicular to the plane containing its other base.

    Of a prism: A line segment drawn from any point of one base of the prism perpendicular to the plane containing its other base.

    Of a pyramid: A line segment drawn from the vertex of the pyramid perpendicular to the plane containing its base.

    Of a trapezoid: A line segment drawn from any point on one base of the trapezoid perpendicular to the other base.

    Of a triangle: A line segment drawn from any vertex of the triangle perpendicular to the line containing its opposite side.

analytical geometry: An approach to geometry in which the points of a figure are represented by coordinates on the Cartesian plane and algebraic methods of reasoning are used to study the figure.

analytical geometric proof: A proof in geometry that employs the coordinate system and algebraic reasoning.

analytical representation of a transformation: The functional notation of a transformation using analytical equations.

analyze: to examine methodically by separating into parts and studying their relationships.

angle: A geometric figure formed by two rays that have a common endpoint.

angle addition postulate: If $\angle ABC$ and $\angle CBD$ are adjacent angles then $\angle ABD = ABC + \angle CBD$.

angle bisector: A ray that divides an angle into two adjacent congruent angles.

angle measure: The number of degrees or radians in an angle.

antecedent: The "if" part of a conditional (if…, then…) statement. (See hypothesis.)

Geometry Made Easy

apothem: A line segment drawn from the center of a regular polygon perpendicular to a side of the polygon.

arc length: The distance on the circumference of a circle from one endpoint of an arc to the other endpoint, measured along the arc.

arc measure: The measure of an arc of the circle in degrees or radians; a unique real number between 0 degrees and 360 degrees or between zero and 2 radians.

arc of a circle: See major arc, minor arc.

area of a polygon: The unique real number assigned to any polygon which indicates the number of non-overlapping square units contained in the polygon's interior.

argument: The communication, in verbal or written form, of the reasoning process that leads to a valid conclusion.

axiom: A statement that is accepted without proof.

axis of symmetry: A line that divides a plane figure into two congruent reflected halves; Any line through a figure such that a point on one side of the line is the same distance to the axis as its corresponding point on the other side.

B

base: Any side or face of a geometric figure to which an altitude is drawn.

betweeness: A point B is between points A and C if and only if $AB + BC = AC$.

biconditional: A statement formed by the conjunction of a conditional statement and its converse; a statement that can be written in "if and only if" form; a definition can always be written as a biconditional statement.

C

Cartesian coordinates: An ordered pair of real numbers that establishes the location or address of a point in a coordinate plane using the distances from two perpendicular intersecting lines called the coordinate axes.

Cartesian plane: The set of all points in a plane designated by their Cartesian coordinates.

center of a dilation: A fixed point in the plane about which all points are expanded or contracted; the only invariant point under dilation.

---

center of gravity: The balance point of an object.

center of a regular polygon: The center of the circle which circumscribes or inscribes a regular polygon.

center of a rotation: A fixed point in the plane about which all points are rotated.

center-radius equation of a circle: The form of the equation of a circle with center (h, k) and radius r given by the formula $(x - h)^2 + (y - k)^2 = r^2$.

central angle: An angle in a circle with vertex at the center of the circle and sides that are radii.

central angle of a regular polygon: An angle in a regular polygon with vertex at the center of the polygon and sides that are radii of its circumcircle.

centroid: The point of concurrency of the medians of a triangle; the center of gravity in a triangle.

chord: A line segment joining any two points on a circle. The diameter is the largest chord of a circle.

circle: The set of all points (or locus of points) in a plane that are a fixed distance, (called the radius) from a fixed point, (called the center).

circumcenter: The center of the circle circumscribed about a polygon; the point that is equidistant from the vertices of any polygon.

circumcircle: A circle that passes through all of the vertices of a polygon. Also called a circumscribed circle.

circumference: The length of or distance around a circle. The formula for circumference is: $C = 2\pi r = \pi d$

clockwise: The direction in which the hands of a clock move around the dial. Used to indicate the orientation of a transformation.

closure: A set "S" and a binary operation "*" are said to exhibit closure if applying the binary operation to any two elements in "S" produces a value that is a member of "S".

collinear points: Points that lie on the same line.

common tangents: Lines that are tangent to two or more circles.

Geometry Made Easy

complementary angles: Two angles the sum of whose measures is 90 degrees.

composition of functions: A way of combining functions in which the output of one function is used as the input of another function; the formation of a new function h from functions f and g using the rule $h(x) = g \circ f(x) = g[f(x)]$ for all x in the domain of for which $f(x)$ is in the domain of g.

compound locus: A set of points satisfying two or more locus conditions.

compound statement: A statement formed from two or more simple statements using the logic connectives, or, and, if...then, or if and only if.

concave polygon: A polygon that has at least one diagonal outside the polygon.

concentric circles: Two or more circles having the same center and different radii.

conclusion: An answer or solution arrived at through logical or mathematical reasoning; the "then" clause in an "if-then" statement; the final statement in a proof which follows logically from previous true statements.

concurrence: The concept of three or more lines intersecting in a single (common) point; having a single point of intersection.

conditional statement: A statement formed from two given statements by connecting them in the form if..., then... .

cone: A solid formed by a circular region (the base) and the surface formed by the straight line segments connecting points on the boundary of the base with a fixed point (the vertex) not in the plane of the base.

conjecture: An educated guess; an unproven hypothesis based on observation, experimentation, data collection, etc.

conjunction: A compound statement formed using the word "and". A conjunction is true only if both clauses are true.

congruent: Having the same size and shape.

conic sections: The plane section created by the intersection of a plane and a cone.

consistency: A property of an axiomatic system where no axiom(s) can be used to contradict any other axiom(s).

constant of proportionality: The number representing the ratio of any two corresponding sides in two similar geometric figures.

construct: To draw a figure using only a compass and a straightedge.

constraints: Any restriction placed on the variables in a problem.

contradiction: A statement that has been shown to be both true and false.

contrapositive of a statement: A statement formed by interchanging the hypothesis and conclusion of a conditional statement and negating each clause.

converse of a statement: A statement formed by interchanging the hypothesis and conclusion of a conditional statement

convex polygon: A polygon is convex if a line segment connecting any two points of the polygon lies entirely in the polygon's interior.

coordinate geometry: An approach to geometry in which a point is represented by coordinates and algebraic methods of reasoning are used; also called analytical geometry.

coordinate plane: The set of all points in a plane designated by their Cartesian coordinates. Also called the Cartesian plane.

coplanar: Any three or more points that lie in the same plane.

corresponding angles: A set of angles formed on each of two or more lines cut by a transversal that are in the same position relative to each line and to the transversal.

corresponding parts: In two geometric figures, the points, sides, and/or angles which are in the same relative position when the figures are placed in a one-to-one correspondence.

counterclockwise: The direction opposite the way in which the hands of a clock move around the dial. Used to indicate the orientation of a rotation.

counterexample: An example that disproves a general statement.

crossection: A plane section perpendicular to the longest axis of a solid.

cube: A polyhedron with six square faces. A cube (or hexahedron) is one of the five platonic solids.

cylinder: A solid geometric figure bounded by two parallel bases which are congruent circles and a lateral surface which consists of the union of all line segments joining points on each of those circles.

D

decagon: A polygon with ten sides.

deductive proof: A formal proof based on logical argument that is justified using axioms and/or theorems.

deductive reasoning: A process of showing that certain statements follow logically from agreed upon assumptions and proven facts; reasoning from the general to the specific.

diagonal: A line segment that connects two non-consecutive vertices of a polygon.

diameter: A chord of the circle that passes through the center of the circle.

diameter of a sphere: A line segment that connects two points on the surface of a sphere and that passes through the center of the sphere.

difference of two perfect squares: A binomial of the form $a^2 - b^2$ which can be factored into $(a - b)(a + b)$.

dihedral angle: An angle formed by two interesting planes.

dilation: A transformation of the plane such that if O is a fixed point, k is a non-zero real number, and P' is the image of point P, then O, P and P' are collinear and $\dfrac{OP'}{OP} = k$.

direct transformation: Any transformation of the plane that preserves orientation.

distance between two points: The length of the line segment joining the two points; a unique non-negative real number.

distance between a point and a line: The length of the perpendicular segment from the point to the line.

distance between two parallel lines: The length of a line segment drawn from any point on one line perpendicular to the second line.

dodecahedron: A polyhedron that has twelve faces. A regular dodecahedron is one of the five Platonic solids and has twelve regular pentagons as faces.

dynamic geometry software: Computer or calculator software used to construct and manipulate geometric figures.

edge of a polyhedron: A line segment that connects two consecutive vertices of a polyhedron.

ellipse: A set of points P_1 in a plane, such that the sum of the distances from P to two fixed points F_1 and F_2 is a given constant k. Any plane section of a circular conical surface which is a closed curve.

endpoint: A point at either the end of a line segment, or arc, or the initial point of a ray.

equiangular: A polygon with all interior angles congruent.

equidistant: At the same distance.

equilateral polygon: A polygon with all sides congruent.

equilateral triangle: A triangle with three congruent sides.

equivalence relation: A relation that exhibits the reflexive, symmetric, and transitive properties.

Euclidean Geometry: The study of geometry based on definitions undefined terms (point, line and plane) and the assumptions of Euclid (c.a. 330 B.C.)

Euclidean Parallel Postulate: Any assumption equivalent to the following statement: If l is any line and P is any point not on l, then there exists exactly one line through P that is parallel to l.

Euler line: For any given triangle, the line that contains the circumcenter, the centroid and the orthocenter.

exterior of a geometric figure: The set of all points outside a geometric figure.

exterior angle of a polygon: An angle formed by a side of a polygon and the extention of an adjacent side.

external segment of a secant: If a secant is drawn to a circle from an external point, the portion of the secant that lies outside the circle.

Geometry Made Easy

F

face of a polyhedron: Any one of the polygons that bound a polyhedron.

fixed point: A point that is its own image under a transformation of the plane.

foot of an altitude: The point of intersection of an altitude and the line or plane to which it is perpendicular.

function: A rule that assigns to each number x in the function's domain a unique number $f(x)$.

G

geometric inequality: A statement in geometry which indicates that quantity is greater than another quantity.

geometric mean: The geometric mean, also called the mean proportional, of two numbers a and b is the square root of their product. If $\frac{a}{m} = \frac{m}{b}$ then m is the geometric mean of a and b.

geometric probability: A probability based on geometric relationships such as area, surface area or volume.

geometric representation of the circular functions: The representation of circular functions on a circle of unit radius. The trigonometric functions are called circular functions because their values are related to the lengths of specific line segments associated with a circle of unit radius.

geometry: Branch of mathematics that deals with the properties, measurement, and relationships of points, lines, angles, surfaces, and solids.

glide reflection: A transformation that is the composition of a line reflection and a translation through a vector parallel to that line of reflection.

golden ratio: When a line segment $\overline{AB}$ is divided by an interior point P such that $\frac{AB}{AP} = \frac{AP}{PB}$, the ratio $\frac{AB}{AP} = \frac{1}{2}(1 + \sqrt{5})$ is called the golden ratio.

golden rectangle: A rectangle whose adjacent sides have a ratio equal to the golden ratio.

graphical representation: A graph or graphs used to model a mathematical relationship.

great circle: The intersection of a sphere with any plane passing through the center of the sphere.

H

half turn: A 180 degree rotation about a point.

hemisphere: Half of a sphere bounded by a great circle.

Heron's formula: The formula expressing the area of a triangle, A, in terms of its sides a, b, and c. $A = \sqrt{s(s-a)(s-b)(s-c)}$ where $s = \frac{1}{2}(a = b = c)$ and is called the semi-perimeter.

hexagon: A polygon with six sides.

hyperbola: Set of points in a plane such that the difference between the distances from P to the foci F_1 and F_2 is a given constant k.

hypotenuse: The side of a right triangle opposite the right angle; the longest side of a right triangle.

hypotenuse and leg triangle congruence: If there exists a one-to-one correspondence between the vertices of two right triangles such that the hypotenuse and leg of one right triangle are congruent to the hypotenuse and corresponding leg of the second right triangle, then the triangles are congruent.

hypothesis: An assumed statement used as a premise in a proof; the "given"; the "if" clause of an "if-then" statement. (See also antecedent.)

icosahedron: A polyhedron having twenty faces. A regular icosahedron is one of the five Platonic solids and has twenty equilateral triangles as faces.

identity elements: For a binary operation * and a set S, I is the identity element if $a * I = a$ and $I * a = a$ for every element a that is in S.

image: The resulting point or set of points under a given transformation; in any function , the image of is the functional value corresponding to .

incenter of a triangle: The center of the circle that is inscribed in a triangle; the point of concurrency of the three angle bisectors of the triangle which is equidistant from the sides of the triangle.

included angle: The interior angle formed by two sides of a polygon.

included side: The side between two consecutive angles in a polygon.

indirect proof : A method of proof in which the statement that is to be proven is assumed false and a contradiction results.

---

inductive reasoning: The process of observing data, recognizing patterns and making generalizations about those patterns.

inscribed angle: An angle whose vertex lies on the circle and whose sides are chords of a circle.

inscribed circle: A circle in the interior of a polygon that is tangent to each side of the polygon.

intercepted arc: An arc of a circle whose endpoints lie on the sides of an angle, and all of the points on the arc are in the interior of the angle.

interior: The set of all points inside a geometric figure.

intersecting lines: Lines that share a common point.

intersection of sets: The intersection of two or more sets is the set of all elements that are common to all of the given sets.

invariant: A figure or property that remains unchanged under a transformation of the plane.

inverse of a statement: A statement formed by negating both the hypothesis and conclusion of a given conditional.

isometry: A transformation of the plane that preserves distance.
If is the image of , and is the image of , then the distance from to is the same as the distance from to .

isosceles trapezoid: A trapezoid in which the non-parallel sides are congruent.

isosceles triangle: A triangle that has at least two congruent sides.

J
There are no J terms in the commencement-level sections of the NYS Mathematics Core Curriculum (Revised March 2005).

K
There are no K terms in the commencement-level sections of the NYS Mathematics Core Curriculum (Revised March 2005).

L
lateral area of a prism: The sum of the areas of the faces of the prism not including the bases.

lateral edge: A line segment that is the intersection of any two lateral faces of a polyhedron.

lateral face: A face of a polyhedron, not including its bases.

length of line segment: The distance between the end two end points of a line segment.

line segment: Given any two points A and B, $\overline{AB}$ is equal to the union of points A, B, and all of those points between A and B.

line symmetry: A geometric figure has line symmetry if the figure is the image of itself under a reflection in a line.

linear pair of angles: Any two adjacent angles whose non-common sides form a line.

locus of points: The set of all points satisfying a given condition or conditions.

logical equivalence: Statements that have the same truth value.

M

major arc: In a circle, any arc whose length is greater than the length of a semicircle.

mean proportional: The mean proportional, also called the geometric mean, of two numbers a and b is the square root of their product. If $\dfrac{a}{m} = \dfrac{m}{b}$ then m $= \sqrt{ab}$ is the geometric mean of a and b.

measure of an arc: The measure of the central angle that subtends the arc.

median of a trapezoid: A line segment that connects the midpoints of the two non-parallel sides of the trapezoid.

median of a triangle: A line segment that connects any vertex of a triangle to the midpoint of the opposite side.

midpoint: A point that divides a line segment into two congruent line segments.

midsegment: A line segment that connects the midpoints of two sides of a triangle; Also called the midline.

minor arc: In a circle, any arc whose length is less than the length of a semicircle.

N

negation: For any given statement , its negation is the statement , $\sim p$ (not p) whose truth value is the opposite of the truth value of p.

net: A two dimensional pattern consisting of polygons which can be folded to form a polyhedron.

n-gon: A polygon with n sides.

non-collinear points: Three or more points that do not lie on the same line.

non-coplanar points: Four or more points that do not lie on the same plane.

non-Euclidean geometry: A geometry that contains an axiom which is equivalent to the negation of the Euclidean parallel postulate.

> **Examples:**
>
> **Riemannian geometry:** A non-Euclidean geometry using as its parallel postulate any statement equivalent to the following: If l is any line and is any point not on l, then there are no lines through P that are parallel to l. (Also called elliptic geometry.)
>
> **hyperbolic geometry:** A non-Euclidean geometry using as its parallel postulate any statement equivalent to the following: If l is any line and P is any point not on l, then there exists at least two lines through P that are parallel to l.

O

Oblique line and a plane: A line and a plane that are neither parallel nor perpendicular.

obtuse angle: An angle whose measure is greater than 90 degrees and less than 180 degrees.

obtuse triangle: A triangle having one obtuse angle.

octagon: A polygon with 8 sides.

octahedron: A polyhedron having eight faces. A regular octahedron is one of the five Platonic solids and has eight equilateral triangles as faces.

one-to-one function: A function where the inverse is also a function.

opposite rays: Two collinear rays whose intersection is exactly one point.

opposite side in a right triangle: The side across from an angle. In a right triangle the hypotenuse is opposite the right angle and each leg is opposite one of the acute angles.

opposite transformation: A transformation of the plane that changes the orientation of a figure.

ordered pair: Two numbers that are used to identify the position of a point in a plane. The two numbers are called coordinates and are represented by (x, y).

ordered triple: Three numbers that are used to identify the position of a point in space. The three numbers are called coordinates and are represented by (x, y, z).

ordinate: The vertical coordinate of a two-dimensional rectangular coordinate system; usually denoted by y.

orientation: The arrangement of the points, relative to one another, after a transformation; the reference made to the direction traversed (clockwise or counterclockwise) when traveling around a geometric figure.

origin: The point in the Cartesian coordinate plane at which the horizontal and vertical axes intersect, designated by the ordered pair $(0, 0)$.

orthocenter: The point of concurrence of the three altitudes of a triangle.

P
parabola: Any plane section of a circular conical surface by a plane parallel to the slant height of the cone.

paragraph proof: A written proof in which the statements and their corresponding reasons are written, in paragraph form, using complete sentences.

parallelepiped: A prism whose bases are parallelograms.

parallel lines: Two or more coplanar lines that do not intersect. Parallel line segments or rays are line segments or rays that are subsets of parallel lines.

parallel planes: Two or more planes that do not intersect.

parallel postulate: Any postulate or axiom that designates the number of lines through a given point that are parallel to a given line.

parallelogram: A quadrilateral in which both pairs of opposite sides are parallel.

parameter: A quantity or constant whose value varies with the circumstances of its application.

pentagon: A polygon with 5 sides.

perimeter: The sum of the lengths of all the sides of any polygon.

perpendicular bisector: A line, segment or ray that is perpendicular to a line segment at its midpoint.

perpendicular lines: Two lines that intersect to form right angles.

perpendicular planes: Two planes that intersect to form right dihedral angles.

pi: The irrational number equal to the length of the circumference of a circle divided by the length of its diameter.

plane: An undefined term in geometry usually visualized as a flat surface with no thickness that extends indefinitely in two dimensions.

Platonic solids: The five regular polyhedra: tetrahedron, cube, octahedron, dodecahedron and icosahedron.

point: An undefined term in geometry usually visualized as a dot representing a non-dimensional location in space.

point of concurrency: A point that is the intersection of three or more lines.

point of tangency: The point where a tangent line intersects a curve.

point-slope equation of a line: The equation of a line formed using its slope and the coordinates of a point on the line, where m is the slope of the line and are the coordinates of the given point.

point symmetry: A geometric figure has point symmetry if every point on the figure is the image of itself under a rotation of 180° about some fixed point.

polygon: A closed plane figure formed by three or more line segments that meet only at their endpoints.

polyhedron: A solid figure bounded by polygons.

position vector: A coordinate vector whose initial point is the origin. Any vector can be expressed as an equivalent position vector by translating the vector so that it originates at the origin.

postulate: A statement assumed to be true without proof.

preimage: The original point or points of a transformation.

premise: A proposition upon which an argument is based or from which a conclusion is drawn.

prism: A polyhedron with two congruent, parallel, polygonal bases and whose lateral faces are parallelograms.

proof: A logical argument that establishes the truth of a statement; a valid argument, expressed in written form, justified by axioms, definitions, and theorems.

proof by contradiction: A method of proof which demonstrates the truth of an implication by proving that the negation of the conclusion of that implication leads to a contradiction; also called an indirect proof.

proportional: Two variables are proportional if they maintain a constant ratio. See also direct variation.

pyramid: A polyhedron having a polygonal base and triangles as lateral faces.

Pythagorean theorem: The mathematical relationship stating that in any right triangle the sum of the squares of the lengths of the two legs is equal to the square of the length of the hypotenuse; if a and b are the lengths of the legs and c is the length of the hypotenuse, then $a^2 + b^2 = c^2$.

Q
quadrant: The four regions of a plane created by the intersection of the coordinate axes.

quadratic equation: An equation that can be written in the form $a^2 + b^2 + c = 0$, where a, b, and c are real constants and $a \neq 0$.

quadrilateral: A polygon with 4 sides.

R
radical: The root of a quantity as indicated by the radical sign.

radius: A line segment drawn from the center of a circle to a point on the circle.

ray: Given any two points A and B, $\overrightarrow{AB}$ is equal to the union of $\overline{AB}$ and all of those points X such that B is between X and A.

reason: A true statement justifying a step in a proof; the use of logic, examples, etc. to determine a result.

rectangle: A parallelogram containing one right angle; a quadrilateral with four right angles.

rectangular coordinates: An ordered pair of real numbers that establishes the location of a point in a coordinate plane using the distances from two perpendicular intersecting lines called the coordinate axes. (See also Cartesian coordinates.)

reflection: An isometry where if l is any line and P is any point not on l, then $r_l(P) = P$ where is the perpendicular bisector of $\overline{PP'}$ and if $P \notin l$ then $r_l(P) = P$.

reflexive property of equality: A property of real numbers that states $a = a$.

regular polygon: A polygon which is both equilateral and equiangular.

regular pyramid: A pyramid all of whose faces are equilateral triangles. Also called a tetrahedron.

remote interior angles: Either interior angle of a triangle that is not adjacent to a given exterior angle of the triangle. Also called non-adjacent interior angles.

restricted domain: The domain resulting from a restriction placed on a function, based on the context of the problem.

rhombus: A parallelogram with two adjacent congruent sides; a quadrilateral with four congruent sides.

right angle: An angle formed by two perpendicular lines, the measure of which is 90°.

right circular cylinder: A cylinder whose bases are circles and whose altitude passes through the center of both bases.

right circular cone: A cone whose base is a circle and whose altitude passes through the center of its base.

right pyramid: A pyramid whose lateral faces are isosceles triangles.

right triangle: A triangle with one right angle.

rotation: An isometry where if P is a fixed point in the plane, θ is any angle and $A \neq P$ then $R_{P,\theta}(A) = A'$ where $m\angle APA' = \theta$ and $R_{P,\theta}(P) = P$.

rotational symmetry: A geometric figure has rotational symmetry if the figure is the image of itself under a rotation about a point through any angle whose measure is not a multiple of 360°.

S

SAS triangle congruence: If there exists a one-to-one correspondence between the vertices of two triangles, such that two sides and the included angle of one triangle are congruent to the corresponding two sides and included angle of the second triangle, then the two triangles are congruent.

SAS Similarity Theorem: If there exists a one-to-one correspondence between the vertices of two triangles, such that two pairs of corresponding sides are proportional and their included angles are congruent, then the two triangles are similar.

scalene triangle: A triangle with no congruent sides.

secant (of a circle): A line that intersects a circle in exactly two points.

secant (of an angle): For a given acute angle θ in a right triangle, sec θ, is the ratio of the length of the hypotenuse to the length of the side adjacent to the acute angle θ; the reciprocal of the cosine ratio of the given angle. See also circular function.

sector of a circle: A region bounded by an arc of the circle and the two radii to the endpoints of the arc.

segment of a circle: The region bounded by a chord and the arc subtended by that chord.

semi-circle: Either of the arcs of a circle determined by the endpoints of a diameter.

set: A well-defined collection of items.

similar polygons: Two polygons which have the same shape but not necessarily the same size.

skew lines: Two non-coplanar lines that do not intersect.

slant height: Of a *pyramid*: The altitude of a lateral face of a pyramid

slope: The measure of the steepness of a line; the ratio of vertical change to horizontal change; if point P is (x_1, y_1) and point Q is (x_2, y_2) the slope of $\overline{PQ}$ is $\dfrac{\Delta y}{\Delta x} = \dfrac{y_2 - y_1}{x_2 - x_1}$.

slope - intercept equation of a line: The equation of a line formed using its slope and its y-intercept. If the coordinates of the y-intercept of the line are $(0, b)$ and the slope is m, then the equation of the line is $y = mx + b$.

sphere: The locus of points in space at a given distance from a fixed point.

square: A rectangle with two congruent adjacent sides.

SSS triangle congruence: If there exists a one-to-one correspondence between the vertices of two triangles, such that all three sides of one triangle are congruent to the corresponding sides of the second triangle, then the two triangles are congruent.

straightedge: An object with no marked units of measure that is used for drawing straight lines

substitution property: Any quantity can be replaced by an equal quantity.

subtraction property of equality: If the same or equal quantities are subtracted from same or equal quantities, then the results are equal.

supplementary angles: Two angles the sum of whose measures is 180 degrees.

surface area: The sum of the areas of all the faces or curved surfaces of a solid figure.

survey: A gathering of facts or opinions by asking people questions through an interview or questionnaire.

symmetric property of equality: A property of the real numbers that states: If then $b = a$.

T

tangent circles (internal): Two circles are internally tangent if they intersect in exactly one point and one circle lies in the interior of the other circle.

tangent circles (external): Two circles are externally tangent if the meet in exactly one point and neither circle has any points in the interior of the other circle.

tangent line to a circle: A line that intersects a circle in exactly one point.

tangent segment: A line segment that is a subset of a tangent line. This usually refers to the line segment drawn from an external point to the point of tangency.

tessellation: A repeating pattern covering a plane.

tetrahedron: A polyhedron with four faces; one of the five Platonic solids that has four equilateral triangles as faces (pyramid).

theorem: A general statement that requires proof.

three-dimensional space: The set of all points in space. The position of each point can be represented by a unique ordered triple (x, y, z).

transformation: A one-to-one mapping of points in the plane to points in the plane.

transformational geometry: A method for studying geometry that illustrates congruence and similarity by the use of transformations.

transformational proof: A proof that employs the use of transformations.

transitive property of equality: A property of the real numbers that states: If $a = b$ and $b = c$ then $a = c$.

translation: A transformation where every point moves the same direction through the same distance.

transversal: A line that intersects two (or more) other lines in distinct points.

trapezoid: A quadrilateral with exactly one pair of parallel sides.

triangle inequality theorem: In any triangle, the sum of the lengths of two sides is greater than the length of the third side.

trichotomy property: A property of the real numbers that states: for every x and y, one and only one of the following conditions is true: $x < y$; $x = y$, $x > y$.

trigonometry of the right triangle: The trigonometric functions for acute angles are the ratios of the sides of the right triangle containing the angle.

truth value: A value, (typically T or F), indicating whether a statement is true or false.

---

two column proof: The outline of a written proof in which the statements and their corresponding reasons are listed in two separate columns.

two-dimensional space: The set of all points in the plane. The position of each point can be represented by a unique ordered pair (x, y). Figures such as angles, pairs of parallel and intersecting lines, circles and polygons exist in two-dimensional space.

U

undefined terms: The fundamental components of an axiomatic system whose understanding is agreed upon but not formally defined. In geometry undefined terms traditionally include point, line, and plane.

union of sets: The union of two or more sets is the set of all elements contained in at least one of the sets.

V

valid argument: A logical argument supported by known facts or assumed axioms; an argument in which the premise leads to a conclusion.

vector: A quantity that has both magnitude and direction; represented geometrically by a directed line segment.

vertex of an angle: The point of intersection of the two rays that form the sides of the angle.

vertex of a polygon: A point where the edges of a polygon intersect.

vertex of a cone or pyramid: The fixed point, not in the plane of the base, to which all points on the perimeter of the base are connected.

vertical angles: The two nonadjacent angles formed when two lines intersect.

volume: A measure of the number of cubic units needed to fill the space inside a solid figure.

X

x-axis: One of the two intersecting lines used to establish the coordinates of points in the Cartesian plane; in that plane, the line whose equation is $y = 0$; in space the axis perpendicular to the yz-plane.

x-coordinate: The first coordinate in any (x, y) ordered pair; the number represents how many units the point is located to the left or right of the y-axis; also called abscissa.

x-intercept: The point at which the graph of a relation intercepts the x-axis. The ordered pair for this point has a value of $y = 0$.

Y

y-axis: One of the two intersecting lines used to establish the coordinates of points in the Cartesian plane; in that plane, the line whose equation is $x = 0$; in space the axis perpendicular to the xz-plane.

y-coordinate: The second coordinate in any (x, y) ordered pair; the number represents how many units the point is located above or below of the x-axis; also called ordinate.

y-intercept: The point at which a graph of a relation intercepts the y-axis. The ordered pair for this point has a value of $x = 0$.

Z

z-axis: A line perpendicular to the plane determined by the x-axis and y-axis at their point of intersection; this axis is used as a reference to determine the third component of the ordered triple (x, y, z).

z-coordinate: The third coordinate in any (x, y, z) ordered triple; the number represents how many units the point is located above or below of the xy-plane.

zero product property: If a and b are real numbers, then $ab = 0$ if and only if $a = 0$ or $b = 0$, or a and $b = 0$.

INDEX

A

Abscissa 65
Acute Triangle 36
Addition Property of Equality 27
Altitude 44, 61
Altitude and Height 26
Analytic Proof Examples 73-74
Angle 24, 29
    Acute Angle 25
    Adjacent Angles 25
    Circles Angles 103-106
        Central Angle 103-104
        External Angle 105
        Inscribed Angle 103-104
        Interior Angle 106
    Complementary Angles 26
    Consecutive Angles 25
    Congruency of Angles 30
    Interior Angle of a Polygon 26
    Obtuse Angle 25
    Opposite Angles 25
    Reflex Angle 25
    Straight Angle 25
    Supplementary Angles 26
    Triangle 36
Angle Bisector 30
Arc 101, 110
    Length 101
    Major Arc 101
    Minor Arc 101
    Semi-circle 101
Area 35, 100
Axis of Symmetry 75

B

Base 26
Basic Loci 96
Betweeness 28
Biconditional Statements 16
Bisect 24
Bisector 28
    Perpendicular Bisector 30, 61
    Angle Bisector 61
    Constructions 91

C

Centroid 63
Circle 100, 100-115
    Angles of a Circle
        Central Angle 103-104
        External Angle 105
        Inscribed Angle 103-104
        Interior Angle 106
    Arc 101, 110
        Length 101
        Major Arc 101
        Minor Arc 101
        Semi-circle 101
    Area 100
    Chord 102, 107, 110
    Circumference 100
    Congruent 110
    Diameter 100, 110
    Equation of a Circle 114
    External Point 108-109
    Radius 100, 110
    Secant 86, 108-109
    Segment 107
    Subtend 100
    Tangent 102, 108-111
    Translation of a Circle 115
Circumcenter 64
Circumscribed Circle 62, 64
Collinear Points 24, 47
Complement 11
Complementary Angles 26
Composite Dilation & Similarity 84
Composite Transformation 85
Compound Locus 95
Concave Polygon 32
Concurrence in Triangles 41-64
Conditional 23
    Hypothesis 23
    Conclusion 23
Cone 8
Congruent 24, 27, 30
    Congruency of Angles 301
        Construction 92
    Congruent Polygon 33
Congruent vs. Equal 27

Geometry Made Easy